THE FEASIBILITY OF JOINT IMPLEMENTATION

ENVIRONMENT & POLICY

VOLUME 3

The Feasibility of Joint Implementation

Edited by

Catrinus J. Jepma

Professor of Economics,
Open University,
University of Groningen,
The Netherlands

SPRINGER-SCIENCE+BUSINESS MEDIA, B.V.

A C.I.P. Catalogue record for this book is available from the Library of Congress.

ISBN 978-90-481-4533-1 ISBN 978-94-015-8559-0 (eBook)
DOI 10.1007/978-94-015-8559-0

Printed on acid-free paper

Table of Contents

Preface and Acknowledgements

This volume contains the papers presented at a conference on 'Joint Implementation' (JI) held on 1-3 June 1994 near Groningen, and organized by IDE-Foundation on behalf of The Netherlands' Ministry of Housing, Spatial Planning and Environment. The main goal of the conference was to evaluate the rationale, potential and feasibility of JI from an academic, policy, and private sector point of view.

The idea to organize a conference on JI somewhat less than a year before the first Conference of the Parties (CoP 1), and shortly before the 10th Session of the Intergovernmental Negotiating Committee for a Framework Convention on Climate Change, originated with Bert Metz and Henk Merkus, both with The Netherlands' Ministry of Housing, Spatial Planning and Environment. During the preparation of the conference and the subsequent preparation of this volume, Henk Merkus proved to be a very competent and helpful contact. His support for the project was invaluable.

The 161 participants as a group provided the intellectual input for the conference, and they should be thanked for their contributions. Special thanks are due to the authors of the papers; they provided the bones and a substantial part of the meat of the conference. All of their contributions have been included in this volume. Similarly, thanks are due to the various commentators. Because of space limitations their contributions have only implicitly been taken into account, insofar as the various contributors have revised their text.

During the editing process the willingness of the contributors to cooperate within the rather strict time frame became clear to me once again, for which I am grateful.

A special word of thanks is due to Wytze van der Gaast. During the conference and its preparatory stage as well as during the editing stage he played a central role behind the scene, taking care of the many facets which made this project possible. It was a pleasure to implement the project jointly with him; he deserves part of the 'credits'. In this respect I may say that the project will get a follow-up in the form of a magazine on JI (the Joint Implementation Quarterly, JIQ, launched in 1995), on which we hope to collaborate further.

Finally, I would like to thank Jan van Ettinger (language check) and Jos Helmich (lay-out) for providing support during the final stage of the preparation of the manuscript.

Abbreviations

ACMAD	African Centre of Meteorological Applications for Development
ADEME	l' Agence de l' Environment et de la Maîtrise Energy
AEEMTRC	ASEN-EC Energy Management Training and Research Centre
AES	Applied Energy Services
ASEAN	Association of South East Asian Nations
BANOBRAS	Mexican National Bank of Public Works and Services
BCSD	Business Council for Sustainable Development
BOF	Basic Oxygen Furnace
C	Carbon
CEE	Central and Eastern Europe
CETES	Government of Mexico Treasury Bills
CEZ	National electricity distribution company in Czech Republic
CFE	Comisión Federal de Electricidad (Federal Electricity Commission)
CFCs	Chlorofluorocarbons
CFL	Compact Fluorescent Light bulbs
CH_4	Methane
CHP	Combined heat and power generation
CIS	Commonwealth of Independent States
CITES	Convention on International Trade in Endangered Species
CNG	Compressed Natural Gas
CO_2	Carbon dioxide
CoP	Conference of the Parties
CRP	Conservation Reserve Program
CTG	Coal-to-Gas
CWA	Clean Water Act
DC	Developing Countries
DG	Directorate General
DSM	Demand Side Management
EAF	Electric Arc Furnace
ECIP	European Community Investment Partners Programme
ECN	Netherlands Energy Research Foundation ECN
EDF	Environmental Defense Fund
EFOM-ENV	Energy Flow Optimization Model for the Environment
EIS	Environmental Impact Statement
EITs	Economies in Transition
ENDA	Environment and Development Action
EPAct	Energy Policy Act
EPZ	Export Processing Zones
ESCAP	Economic and Social Commission for Asia and the Pacific
ETB	Energy Tool Box
ETW	Energy for Tomorrow's World
EU	European Union
FACE	Forests Absorbing Carbon dioxide Emissions
FCCC	Framework Convention on Climate Change
FFI	Forests for the Future

FIDE	Mexican Trust Fund for the Program of Energy Conservation of the Power Sector
FMO	Netherlands Finance for Development Company
GDP	Gross Domestic Product
GEF	Global Environment Facility
GET	Global Environment Trust Fund
GHG	Greenhouse Gas(es)
GJ	Gigajoules
GtC	Gigatons of Carbon
Gtoe	Gigatons oil equivalent
GWP	Global Warming Potential index
ha	Hectare
HCFC	Halo Chlorofluorocarbons
Ic	Incremental costs
IC	Industrialized Countries
ICB	International competitive bidding
ICI-AMV	Steam reforming process (AMV) in Imperial Chemical Industry (England)
IDB	Incremental domestic benefits
IDE	Foundation for International Development Economics
IEA	International Energy Agency
IEI	Instituto de Economia Industrial
IFC	International Finance Corporation
IIASA	International Institute for Applied Systems Analysis
INC	Intergovernmental Negotiating Committee
INE	Instituto Nacional de Ecologia (Mexican National Environmental Institute)
IOU	Investor Owned Utilities
IPCC	Intergovernmental Panel on Climate Change
IPP	Independent power producer
IPTA	Investment Promotion and Technical Assistance
IRR	Internal Rate of Return
JI	Joint Implementation
Kg	Kilogramme
Kgoe	Kilogramme oil equivalent
Km	Kilometre
Ktoe	Kilotons oil equivalent
kWh	Kilo Watt hour
LLB	Lawrence Berkeley Laboratory
LRMC	Long-run marginal costs
Mtoe	Mega ton oil equivalent
Mha	Mega hectares
MJ	Mega Joules
MW	Mega Watt
NGOs	Non-Governmental Organizations
NIEO	New International Economic Order
NO_x	Nitrogen oxide

NOP-MLK	Netherlands National Research Programme on Global Air Pollution and Climate Change
NSJI	North-South JI
NYSPSC	New York State Public Service Commission
ODA	Official Development Aid
OECD	Organization for Economic Cooperation and Development
OPEC	Organization of Petroleum Exporting Countries
PCU	Project Coordination Unit
PTA	Preferential Trade Area
SEC	Specific Energy Consumption
SEDESOL	Secretaria de Desarrollo Social (Mexican Ministry of Social Development)
SEDUE	Mexican (ex-) Ministry of Urban Development and Ecology
SEP	Samenwerkende Electriciteits Producenten (The Netherlands' electricity generating board)
SEQRA	State Environmental Quality Review Act
SO_2	Sulphur dioxide
SSJI	South-South JI
STAP	Scientific and Technical Advisory Panel
t	Ton
T	Transfer
TERI	Tata Energy Research Institute
TFAP	Tropical Forest Action Plan
toe	Ton oil equivalent
TWh	Tera Watt hour
UN	United Nations
UNCED	United Nations Conference on Environment and Development
UNEP	United Nations Environment Programme
UNIDO	United Nations Industrial Development Programme
UN ECE	United Nations Economic Commission for Europe
UN ECA	United Nations Economic Commission for Africa
US CAP	United States Climate Action Plan
USA	United States of America
USIJI	United States Initiative on Joint Implementation
USSR	Union of Soviet Socialist Republics
WCED	World Commission on Environment and Development
WEC	World Energy Council

Introduction

CATRINUS J. JEPMA

This volume contains the various contributions that were made during the International Conference on Joint Implementation, held near Groningen, The Netherlands, 1-3 June 1994. The conference was initiated by The Netherlands' Ministry of Housing, Spatial Planning and Environment - along with the Ministries of Foreign (Development Cooperation) and Economic Affairs - and carried out under the responsibility of Foundation IDE (Groningen, The Netherlands). Its underlying idea was to bring together an international group of specialists on Joint Implementation (JI) - from governments, NGOs, business and science - to discuss its feasibility of JI. The conference was timed between the 9th and 10th session of the Intergovernmental Negotiating Committee (INC) for a Framework Convention on Climate Change (FCCC) - which was by then considered as the most crucial phase preparing for the first Conference of Parties (CoP 1; Berlin, 28 March - 7 April 1995) - in order to provide additional information or results that might serve as an input in the international negotiating process.

The main purpose of the conference was, however, to sit back and have a reflection about what has now become known as JI, and evaluate from the perspective of academics, and practitioners, jointly with officials what promise the option really holds, and to evaluate under what socio-economic and political circumstances and conditions one could successfully proceed in setting out the framework for its further testing and application.[1] Recognizing that JI is not just an academic concept or just a practical tool, but also a key topic of West-East-South discussions, during the conference a rather spontaneous process started to explore whether the participants - 161 specialists from 43 countries, 37 of which were from 20 developing countries and 28 from 9 countries in transition - could reach consensus about a text that would reflect the spirit of the conference. This finally led to what has become known as the Groningen Statement on Joint Implementation (see p. 8), which was distributed widely after the conference.

The many contributions during the conference have been organized in this volume around the following four topics:

- JI as a concept (Part I);
- JI and its institutional setting (Part II);

[1] The conference was combined with a lead authors' meeting of IPCC WG III, chapter 7; the contributions of Jefferson, Jepma/Lee, Maya, Asaduzzaman, and Sedjo (ch. 2, 3, 14, 19, 22) served the dual purpose of providing an input into the conference and acting as a discussion paper for IPCC Chapter 7 'Generic Assessment of Response Options'.

C. J. Jepma (ed.), The Feasibility of Joint Implementation, 1–8.

- JI from the policy perspective (Part III); and
- JI and practical applications (Part IV).

Part I

In the first part, the fundamental issue is raised what the benefits of the option could be, and in particular, how these benefits are to be distributed among the parties involved. The overall perspective, as indicated by Pearce, seems to be that JI's major promise is that it can serve to improve the cost-effectiveness of meeting internationally agreed emissions reduction targets and as a vehicle for international financial and technology transfers. Thus it can potentially provide a net overall gain for all parties involved, which is probably why the concept has been adopted also in other environmental treaties (Pearce, pp. 16-8). Obviously, the concept presupposes sufficiently large international differences in marginal greenhouse gas (GHG) abatement costs, which are not overtaken by the transaction costs involved with operationalizing the option. However, on the whole one seemed to be rather optimistic about that overall net positive welfare potential (for some simulation outcomes, see Jepma and Lee, pp. 60-1, and Swisher and Villavicencio, pp. 259-61).

Having said this, the issue becomes under what circumstances it can really become a 'win-win-win' instrument, *i.e.* the climate wins, as well as both parties involved. The main concerns expressed in particular in ór on the industrialized countries are that JI may provide a means to postpone necessary adjustments in production, consumption and lifestyle at home; the 'recipient' countries from their part have expressed concerns that the funding will turn out to be not sufficiently additional, or that the industrialized countries are unloading their greenhouse problem on them. In addition, the various parties involved in projects in the spirit of the JI-philosophy are concerned about the possibility that the discussions will (by fully involving non-signatories in the negotiations, Pearce p. 27) take an unnecessary long time and will become increasingly complicated. The present enthusiasm to actually start (*e.g.* through private sector carbon offsets; for an overview, see Pearce, p. 23) may thus disappear, as well as JI's advantage of offering a means of early action (stressed by Jefferson, p. 53).

All this means that JI has to be designed and thought through carefully, whereby one faces the dilemma that, by pushing the option too hard or not hard enough, resistance may increase or interest may fade away respectively. This calls for a balanced approach and a debate, in which potential benefits, costs and risks are carefully evaluated at various aggregation levels (Vellinga and Heintz, pp. 71-3), and in which many of the uncertainties have to be clarified before one actually can proceed (Hare and Stevens, pp.

79-85). It also calls for a fruitful North-South dialogue, in which both parties are prepared to carefully listen to each other and develop a common language (see contribution of Yaker, pp. 87-97).

Part II

Given the general reflections about the pros and cons of JI, and the main lines along which its discussion is progressing, the question arises how the institutional mechanisms should be designed to make JI actually work. JI can be seen as an outcome of a process in which supply and demand have to match. On the one hand, one party - commonly an industrialized country - is looking for a cost-effective option outside its own territory; on the other hand there is a party - a country in transition or a developing country - which believes that it can provide such a cost-effective opportunity. If supply meets demand, a JI project can result. But how does the allocation process look like and, more specifically, how can one prevent the transaction costs from becoming a prohibitive obstacle?

In the contributions to Part II of this volume, various practical complications were noticed, such as: to determine the proper baseline so as to assess the JI net contribution; the risk of double-counting; strategic behaviour anticipating JI actions; the division of responsibilities between the private and public parties involved; monitoring and control procedures and requirements; and last but not least, the distribution of the benefits between both parties.

One seemed to agree that a multi-facetted, and carefully designed pilot phase might provide the most promising road to proceed and that one should beforehand be clear about the economic and legal criteria for JI (Spradley, Chapter 7; Schrijver, Chapter 9). Some creative thinking on how the aforementioned supply and demand might match is presented by Wrexler et al. (Chapter 8), who outline various stages of intermediation in 'a phased approach to evolving the market architecture of the JI regime'. Clearly, various questions of how a pilot stage might precisely look like still have to be answered.

An example of different points of view on the timing of the acceptance of JI are Vellinga and Heintz (pp. 143-7) on the one hand, and Helme and Gille (pp. 147-160) on the other hand. Vellinga and Heintz take a reserved position by arguing that one should follow a phased approach during which, after a pilot phase, a second transition phase should be designed in which countries could adopt dual (*i.e.* domestic and JI) and non-interchangeable commitments, before one would enter the final phase of full JI application. Crediting would in their view have to be postponed rather far into the future. Helme and Gille from their part argue that a JI system will fail to come off the ground if the moment at which parties will be remunerated somehow for their JI investments is shifted too much to future. They are, instead, in favour of moving to the final phase right after the pilot phase, and of already experimenting in an early stage with bilateral agreements at the project level, such as those established between the USA and some Czech and Slovak municipalities.

Part III

Part III of the volume puts the policy makers' perspective central: they face the dilemma mentioned earlier to reconcile the positions and opinions of the various parties involved.

Almost all participants agreed about the need of proper criteria, also during a pilot phase, and of clear procedures; in addition, various ideas and proposals were put forward.

Metz launches the idea to create a multilateral fund to control the balance of JI actions between industrialized and developing countries (p. 173). Graham sets out the USA view on what the FCCC permits, and more specifically, the criteria JI projects should meet. In this respect it is being argued that the operationalization of the criteria be left to the countries themselves, and that the private sector is the most efficient mechanism for implementing individual projects. Graham also extensively and very clearly sets out the USA Initiative on JI (USIJI), presented at the 9th session of INC (pp. 181-7). Wisaksono illustrates various ASEAN energy use scenarios and how they relate to the energy policies in the various countries in the region (pp. 190-6).

On the basis of a critical perspective, Maya develops various general criteria for North-South JI that should in his view be satisfied for JI to be acceptable to the South (Chapter 14).

Faragö and Nemes express their views and concerns with respect to JI from a country-in-transition point of view, and make a plea for an extensive pilot phase to further evaluate the option and its possible application (Chapter 15).

Verdugo and Leslie, finally, mention a number of carbon offsets and initiatives similar to JI under development or already operational (pp. 224-6), and make a plea for national policies to gather experience, preferably based on a voluntary framework in which the commitments of all parties involved - business, NGOs, the academic community and government - are based on a common concern as regards global climate change (pp. 226-9).

Part IV

Part IV of this volume focuses on the practical application of projects in the sphere of carbon offsets and/or initiatives that might qualify for JI. Both actual projects (some of which were mentioned earlier in Pearce, pp. 25-7, and in Verdugo and Leslie, pp. 224-6) and the potential for projects are discussed.

First, Trexler reviews a series of about 10-15 ongoing offset projects, mainly in the sphere of forest protection and management, and of reforestation (pp. 236-41). He explains why so many of the early projects have been in the forestry sector, and draws a number of preliminary conclusions for private sector participants (pp. 242-3). One of the main general conclusions is that - especially in a voluntary emissions control system - carbon offset projects should work with a limited number of key objectives. If, due to a complicated negotiation process JI projects would have to undergo multiple screening and satisfy several objectives ànd constituencies, *etc.*, funding - he believes - might well become hard to find. This is particularly so if one expects, as Trexler does, that in the short to medium term voluntary private sector funding will dominate (p. 246).

Swisher and Villavicencio (pp. 249-56) discuss a UNEP Abatement Costing Project initiated in 1991, which puts the national abatement cost functions - derived for seven developing and three industrialized countries for various detailed technical options, both in energy use and supply - into an inter-country comparative framework. On this basis, country-specific reference scenarios are subsequently designed for the period up until 2030. The results of the cost functions indicate considerable scope for low cost, and perhaps negative cost options, not only in developing but also in industrialized countries. Their data therefore suggest considerable scope for JI: some quarter of the industrialized countries' medium-term commitments, and increasing as times goes on (results comparable to the results of Jepma and Lee, pp. 60-1, which were based on much more aggregate cost functions). Their data also suggests significant resulting overall cost reductions (pp. 259-61): through JI the average greenhouse abatement cost would in the long run come down from $25/t$CO_2$ to about $10/t$CO_2$.

Asaduzzaman (Chapter 9) surveys energy savings potentials and constraints on an inter-country comparison basis. He argues that a substantial scope for energy efficiency improvement exists, but that it would be too simple - especially in developing countries - to combat market failures just by getting prices right; more information, public education and awareness are equally important.

Blok et al. (Chapter 20) argue in a similar vein by pointing out numerically, on the basis of sector/country data, how common methodologies should be designed to determine the scope for energy efficiency in various sectors/countries (as a potential input in the baseline discussion, editor's note).

Janssen and Van der Vleuten (Chapter 21) focus on the cement industry - in 1990 responsible for some 4% of global anthropogenic CO_2 emissions. They argue that for various reasons this industry could be a suitable candidate for targeting JI, especially if carried out in a North-South context.

Sedjo (Chapter 22) focuses his contribution on forestry, and argues that the information about the costs of this option has increased considerably by designing cost functions rather than single point estimates, and by better taking into account opportunity costs, ancillary effects, economies of scale and discounting; this leads to (upward sloping) marginal cost curves. Recognizing the additional scope for carbon sequestration through forest management, he concludes that on the basis of earlier data the potential of carbon sequestration through tree planting was probably underestimated.

Verweij (Chapter 23) sets out the initiative taken by The Netherlands' electricity producers to create the absorbing capacity (through planting 150.000 ha. of forests in various countries) for a 600 MW modern coal-fired power plant (the so-called FACE project). Contracts have already been signed for this purpose, taking cooperation with local authorities and peoples on the basis of equality as a key criterion.

Wilms (Chapter 24) stresses the potential of the use of Compact Fluorescent Light bulbs (CFLs) in Demand Side Management (DSM) to improve energy efficiency. Experience from various countries shows that consumers can be convinced of the significant saving opportunities of the use of CFLs via utility campaigns. According to Wilms, JI can provide a vehicle to adjust people's behaviour by actually demonstrating the CFLs' advantages: saving on electricity use, peak load decrease, and carbon emissions reduction.

Luzuriaga (Chapter 25) highlights the $23 mln. ILUMEX project, *i.e.* to replace 1.7 mln. residential light bulbs with CFLs in two Mexican cities (editor's note: at the beginning of 1995 the actual implementation of the project, however, has not yet started), including the carefully designed procedures surrounding the project. The project is meant to serve as a replicable model for DSM in the developing countries; it is recognized that a significant part of the project benefits have to be shared with the users (through a rebate) if the project is going to be successful.

Roland and Haugland (Chapter 26) critically assess both the ILUMEX project mentioned and a $44 mln. coal-to-gas conversion project in Poland. They argue that the baseline discussion is rather complex if one takes into account fungibility effects, such as: free riding (investment in measures to reduce emissions would also have been taken without JI); the effect of national/local environmental regulations (measures with similar effects as JI projects are to be implemented sooner or later anyhow); or conservation rebounds (efficiency improvements trigger behaviourial changes - *e.g.* cheaper but more lighting - partly offsetting savings). If one of the parties is not committed to emission reduction, it is actually in both parties' interest to overreport emission reductions. They, therefore, warn for overly optimistic pictures of net abatement of North-South or North-East JI projects and call for a strong verification and monitoring mechanism.

Sanghi et al. (pp. 370-6), finally, point out that some States of the USA (notably New York State), in their environmental regulations, have already introduced provisions for offset requirements, which provisions could result in a JI project. Moreover, externality adders (*i.e.* calculations expressing aggregate emissions of pollutants) are now being used to select from various alternative resources on the basis of resource costs after incorporation of externalities. Sanghi et al. suggest, however, on the basis of some experiences, that the use of the calculated externalities should be taken a step further. Utilities that do not choose to invest in the least cost (including externalities) resource plant could be forced to make expenditures, *e.g.* through JI projects, in order to compensate, say 10%, of the imputed externality costs or corresponding tons of CO_2. According to them, this might create a market for JI projects of more than $1 bn. in the next 15 to 20 year!

Final remark

In conclusion, JI has become one of the key topics in the international policy debate on energy and climate change policies. Early experiences show many practical obstacles and uncertainties but also considerable scope for 'win-win-win' situations, *i.e.* the 'climate' wins, but on the basis of increased cost efficiency of climate change policies, also the investing and host country participant. It seems that the most difficult questions still to be answered are how large precisely these gains can be and how the distribution of the gains should look like in order to establish a fruitful dialogue between the various parties. One may hope that a successfully implemented pilot phase will provide the necessary answers to proceed to JI.

GRONINGEN STATEMENT ON JOINT IMPLEMENTATION
3 June 1994

The International Conference on Joint Implementation (JI) held 1-3 June 1994 near Groningen, The Netherlands, brought together more than 160 public officials, from government agencies as well as from international organisations, and representatives from science, industry and NGOs from 43 industrialized and developing countries from all continents of the world. 124 participants came from industrialized countries, including 28 from countries in a state of transition, and 37 came from developing countries; all of them attended in their personal capacity.

JI refers to a form of North-North, North-South and South-South cooperation between two or more Parties to the Framework Convention on Climate Change, for jointly pursuing action designed to reduce or absorb emissions of greenhouse gases, to fulfil parts of the obligations under the Convention. The cooperation should be on a voluntary basis and respect the sovereignty of all States.

During three days, the Conference discussed economic, developmental, environmental, legal and institutional aspects of JI, to a large extent on the basis of plans for or experience with projects which aim to qualify as JI. These projects suggested an important potential of JI, as one of the possibilities at the national and international levels to achieve the objective of the Convention in a cost-effective manner.

At the end of these discussions there was considerable agreement that:

- due to the urgency of the global situation, JI should be a supplement to domestic action by developed Parties to reduce or absorb emissions;
- JI projects would have to follow objective and verifiable criteria (through monitoring) in order to qualify for eventually crediting actual reductions to the cooperating Parties;
- these criteria, such as baselines, and a system of eventually crediting the investing and receiving Parties are still to be established by the Conference of the Parties, the Convention's supreme body;
- in the meantime, however, international cooperation to reduce or absorb emissions of greenhouse gases should be encouraged to gain further experience with the merits and demerits of JI;
- the experience to be gained in this way should serve to gain insight and build confidence in the actual potential of JI and thus facilitate the intergovernmental negotiations still needed.

The above Statement can only give an overall impression of the spirit prevailing at the Conference. Its full report should, in due time, give a balanced account of the intensive discussions of the main aspects of JI, including the concerns with North-South JI expressed especially by participants from developing countries and NGOs. In designing JI projects with developing countries and countries in transition, their priorities should be taken into account.

In addition to this report, the participants recommended the organizers to explore the possibilities of efficient ways to sustain the process started by the Conference, such as a JI newsletter and a network for the exchange of information on JI studies and projects. It was also recommended that a follow-up conference should be held in a developing country focusing on JI from the perspectives of developing countries. A similar conference was recommended for countries in transition.

Finally, the participants expressed their warm appreciation to The Netherlands' government for inspiring and enabling the Foundation IDE to organize this most hospitable and productive Groningen Conference.

Opening Address to the Conference

The Netherlands' Minister of Housing, Spatial Planning and the Environment

HANS ALDERS

Almost five years ago the concept of joint implementation, at the time a concept 'avant la lettre', was introduced at the international Climate Conference in Noordwijk in 1989. For that meeting, a report had been drawn up, underlining the need for an international cost-effective approach to the climate issue. The key assumption was and still is that the marginal costs of measures, aiming at reducing emissions of greenhouse gases, could differ drastically from country to country. This played a major role in the negotiations which resulted in a United Nations Framework Convention on Climate Change, presented exactly two years ago, at the UN Conference on Environment and Development.

One of the most imported passages, which helped to ensure that many countries supported the Convention, was the one in which it was stipulated that the Parties could comply with their obligations individually, but also jointly. This concept has come to be known as 'joint implementation'. Few people however realized how much research, development, information, persuasion and negotiations would be needed to bring this part of the Convention closer to practice. The Convention itself simply says that the criteria for applying Joint Implementation have to be decided on at the first meeting of the Conference of the Parties. However, the Convention has been in force for more than ten weeks and many countries are busy complying with their reporting obligation. In March 1995 the parties will meet for the First Conference of the Parties. Various countries, including The Netherlands, are actively elaborating this concept in the Convention in greater detail.

In 1993, The Netherlands' Government set up a study into the legal and economic aspects of joint implementation. The Netherlands was also one of the sponsors of a fruitful international workshop on joint implementation in Southampton (Bermuda) in January this year.

Internationally, the further elaboration of joint implementation commenced during the eighth round of the Intergovernmental Negotiating Committee (INC) in August last year. It was found that the subject of joint implementation was a politically sensitive issue, notably in developing countries. The latter are afraid that the developed countries want to comply with their Convention commitments too cheaply. Moreover, among the group of developed countries, there are different ideas as to what joint implementation really means. The option offers opportunities, but at the same time entails risks. Actually, it does not look as if joint implementation will be used to achieve current commitments under the Convention. I believe that this is right, because otherwise this would convey

C. J. Jepma (ed.), The Feasibility of Joint Implementation, 9–11.

the wrong message, in a historical perspective. It would be unacceptable if the developed countries, thanks to joint implementation, could continue their current CO_2 emissions. The solution to the greenhouse problem requires a radical change in production and consumption patterns, especially in developed countries!

However, the gravity and the scale of the climate problem make it clear that we need all possible instruments and thus that we have to go further with joint implementation. Details have to be worked out, thereby minimising the risks and maximising the benefits. The Netherlands sees joint implementation as a way of combining a cost-effective approach to the climate issue with the launching of new flows of funds in the direction of developing countries and the promotion of the transfer of environmentally-friendly technologies.

Joint implementation funds will have to contribute to projects, which have high priority in developing countries, such as improving electricity generation and abating local air pollution. It is immensely important that joint implementation projects really result in significant benefits for developing countries. Moreover, joint implementation can help to promote a dynamic international approach to the greenhouse problem. It can provide an incentive to developed countries to accept far-reaching obligations under the Climate Convention for the post-2000 period. It can also help us fulfilling the ultimate target of the Climate Convention. The present Convention commitments are most probably not adequate for realizing this target, so this means that we have to reach agreement on stricter targets for the Annex-I Parties. Taking these facts into account, the possibility of achieving some of these commitments by means of joint implementation may come to play an important role.

I am glad that there is general support for the view that joint implementation funds should be clearly seen as additional. They should also not be deployed at the expense of development aid. There also seems to be agreement about a pilot period during which provisional criteria would apply and there would be no question of crediting emission reductions obtained abroad against current Convention commitments. That period could last from three to four years. After thorough evaluation, the Conference of the Parties would be able to take a final decision on the application of joint implementation.

Important criteria, which should be tested during the pilot period, would be above all the possibilities of quantifying emission reductions deriving from joint implementation projects. Other crucial points are the monitoring and the enforcement. Lastly, it has to be ascertained what local benefits derive from the projects. Naturally, it is up to the first Conference of the Parties to decide on the pilot period and the accompanying guidelines for the Parties.

The Netherlands is currently preoccupied with developing its pilot projects' programme. Various ministries and a number of companies and organizations in the community are involved in the process. This programme should be officially presented at the first Conference of the Parties. The Netherlands' Government is also encouraging a national debate on joint implementation. It will be publishing a discussion document very soon and will be asking for comments from interested parties and advice from the major advisory bodies in the field of socio-economic policy, the environment and development cooperation. The results of this debate will help the Government decide on its final viewpoint with respect to the first Conference of the Parties early next year.

This conference here in Groningen makes a vital contribution to the international dialogue, between policy-makers but also between and with the many other participants from the private sector, the research world and environment and development organizations. I would like to compliment the Foundation International Development Economics for the initiative to organize this conference. The ultimate result of the conference will be determined by you, yourselves, in the days ahead. I expect that we will be able to learn a lot from each other in the next three days. It is an honour for me as host to be able to declare this conference open. I wish you all extremely fruitful discussions and I hope that it will be the start of many jointly implemented activities.

PART I

THE SCOPE FOR JOINT IMPLEMENTATION

1. Joint Implementation

A general overview

DAVID PEARCE*

1 Joint Implementation: The Concept

In the context of international environmental agreements, Joint Implementation (JI) involves a bilateral deal, or even a multilateral one, in which countries with high costs of pollution abatement or environmental conservation invest in abatement or conservation in a country with lower costs, and receive credit for the resulting reduction in emissions or increase in conservation. While JI is potentially applicable to any environmental objective, it is generally applied in contexts where one of the partners in the deal has a commitment to reduce pollution emissions. A constraint on the trade is that emission reductions in the low cost country, the 'host' country, must at least offset the avoided reductions in the 'donor' country. The obvious potential attraction of JI is that it reduces the global costs of meeting internationally agreed emission targets. It therefore contributes to cost-minimization. This is fairly self evident: if the donor avoids cutting emissions of X tons at cost C, and invests in cutting emissions in the host nation by X tons at cost αC ($\alpha<1$), then there are cost savings of $(1-\alpha)C$ and no worsening of global environmental quality. However, the latter result, that global quality does not decline, can be guaranteed only if the obligation being traded is 'uniformly mixed', *i.e.* the damage being done does not vary with the location of the bargaining parties. Greenhouse gases (GHG) are examples of such uniformly mixed pollutants: it does not matter where the reduction takes place since 1 ton of a GHG does the same amount of global damage wherever the reduction takes place.

Once JI is extended to other pollutants and environmental damages, the guarantee of non-declining global quality cannot be provided without further restrictions on the trades. For example, acidifying pollutants, such as SO_2 and NO_x, are not uniformly mixed pollutants. Thus a trade between one donor and one host could affect air quality and acidifying depositions in a third party country. Although it has not generally been entertained, trades could be undertaken between countries in conserving an environmental asset such as biological diversity. In practice, the 'value' of biological diversity is likely to be very site specific and no satisfactory 'metric' exists yet for comparing

* David Pearce is director of the Centre for Social and Economic Research on the Global Environment (CSERGE) and professor at University College London and University of East Anglia.

C. J. Jepma (ed.), The Feasibility of Joint Implementation, 15–31.

reductions in different locations.[1] Were such relative values, or 'exchange rates', to be developed and accepted, then JI could be applied in that context also.

Apart from cost savings, JI is widely held to be consistent with other objectives. First, JI could permit national emission targets to be more ambitious than the obligations under any agreement. This outcome is less predictable for sulphur trading, as Klaassen has shown:[2] cost saving may well be at the expense of ecosystem protection. Second, JI can also permit a wider participation in any agreement. In the GHG context this may be very important since it can reduce the 'carbon leakage' problem whereby measures to reduce emissions among one set of countries result in increased GHG emissions in other countries who are not party to the agreement. This can come about because some measures, such as carbon taxes, raise the tax-inclusive price in the participating countries but depress the tax-exclusive price in other countries.[3]

In principle, then, JI can be applied to any international deal in which one or more of the parties has emission targets, or, more generally still, to any international agreement in which the parties have obligations to achieve some environmental goal. Nor is it necessary for there to be an international agreement: country A may trade with country B simply to achieve its own domestic environmental target. In general, however, discussion has focused on JI in the context of international agreements.

One way of thinking about JI is that it is the 'thin end of the wedge' towards a full tradeable obligations scheme. In initial form, it is likely to involve trades in which just a few countries are involved, one donor and one or more hosts, but it is easy to see how it could be developed to embrace a wholesale trading system such as that being developed in the USA under the Clean Air Act amendments of 1990.[4] Most commentators appear to be agreed that the prospects for such a fully integrated global trading system are some way into the future. Others, such as Bohm, have questioned whether JI is a credible precursor to a full traded system.[5]

2 JI in International Agreements

JI has found a firm place in a number of recent international agreements.

---

1 See Dinerstein and Wickramanayake, 1993.

2 Klaassen, 1994a.

3 Bohm, 1994a.

4 Sorrell, 1994; Rico, 1994.

5 Bohm, 1993.

2.1 The Montreal Protocol

JI is enabled under the Montreal Protocol on the control of chlorofluorocarbons (CFCs). Article 2.5 of the 1990 Revisions to the Protocol states:

> 'Any Party may, for any one or more control periods, transfer to another Party any portion of its calculated level of production [of CFCs], provided that the total combined calculated levels of production of the Parties concerned for any group of controlled substances does not exceed the production limits set out in those articles for that group',

thus permitting trade in production quotas.

Article 2.8(a), which deals with the EU, states that:

> 'Any Parties which are Member States of a regional economic integration organization .... may agree that they shall jointly fulfil their obligations respecting consumption under this article provided that their total combined level of consumption does not exceed the levels required by this Article',

effectively placing a 'bubble' over the EC.

2.2 The framework convention on climate change

Article 4.2(a) of the Framework Convention on Climate Change (FCCC) states that

> '.... developed country Parties and other Parties included in Annex I may implement .... policies and measures jointly with other Parties and may assist other Parties in contributing to the objective of the Convention ....'.

The FCCC recognizes different circumstances and capabilities of Parties, and distinguishes the following groups: the Annex I countries (developed countries and those with economies in transition *i.e.* former East Bloc countries), Annex II countries (OECD countries only) and non-Annex countries (developing countries). The Annex I countries have generally recognized, if vaguely worded, obligations to cut emissions of GHG to their 1990 levels by the year 2000. The non-Annex countries have more general obligations.

JI is not defined in the Convention, along with another and related critical economic concept - incremental cost and the criteria for determining JI are to be established by the Conference of Parties at their first meeting early in 1995. Until then the Intergovernmental Negotiating Committee (INC) is the forum for discussions about the meaning of JI in the context of the FCCC.

JI takes on a wider meaning in the FCCC since it may also be the foundation for more general global partnerships in achieving sustainable development. In what follows, we narrow the focus to obligations to control GHG emissions.

2.3 The second sulphur protocol

It is perhaps less well appreciated that JI is also enabled under the Second Sulphur Protocol of the 1979 Convention on Long Range Transboundary Air Pollution, due for signature in Oslo, June 1994. The Protocol covers the UN ECE region, *i.e.* Europe and the economies in transition. Article 2.7 states:

> 'The Parties to this Protocol may, ... in accordance with rules and conditions which the Executive Body shall elaborate and adopt, decide whether two or more Parties may jointly implement the obligations set out in Annex II. These rules and conditions shall ensure the fulfilment of the obligations set out in paragraph 2 .... and also promote the achievement of the environmental objectives set out in paragraph 1 ....'.

Annex II of the Protocol sets out the targets for SOx emission reductions in each country that is party to the Protocol, and which targets are derived from computerized models of emissions, transportation and deposition in the UN ECE area. The inclusion of JI is interesting since the models on which the Protocol is based are cost-minimizing models. Given optimal achievement of the resulting emission targets, which vary by emitting source, JI would have no role to play. Its inclusion in the Protocol is a recognition of the need to ensure cost-efficiency in a context where (a) actual targets will diverge from the optimized ones, and (b) emissions, depositions and costs are the subject of remaining uncertainties.

2.4 The Rhine 'salts' treaty

A 1993 Protocol to the 1976 Convention on the Protection of the Rhine Against Pollution by Chlorides enables the Rhine riparian states to engage in joint funding of downstream measures in The Netherlands to divert chlorides to the North Sea rather than requiring upstream polluters to remove them at source.

3 Cost Efficiency

Just how far JI will result in significant cost savings is difficult to say. Self-evidently, cost savings arise because the marginal costs of abatement are assumed to vary between host and donor nations. But it is not clear how safe this assumption is.

Figure 1.1 shows the most elementary case of JI. The marginal abatement cost curve of the host country is shown as MACh, and that of the donor is shown as MACd. MACh is read from left to right and MACd from right to left. The obligation of the donor is to reduce emissions by OX. In order to minimize costs, OY is reduced in the donor country and YX in the host country at a cost of OYZ + YZX. This is compared to a cost of OWX if all the abatement is concentrated in the donor country, so that the overall cost savings are XZW.

Figure 1.1. Cost Savings in an Elementary Case of JI.

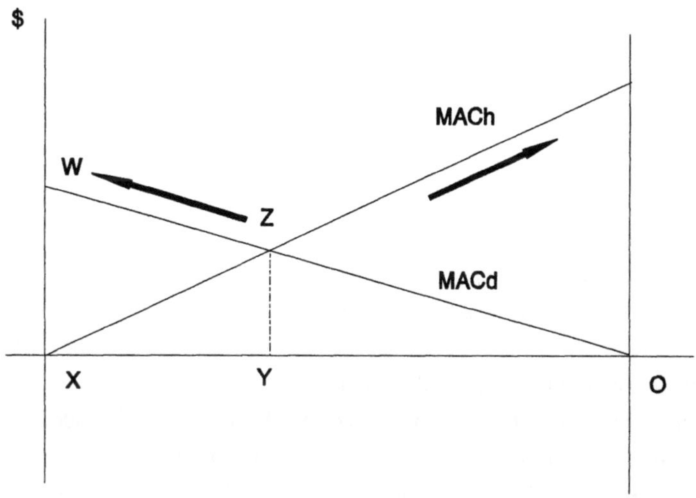

4 Actual Trades

As it happens, a number of trades in carbon have already taken place, so that a modest global carbon market already exists. We distinguish between private and public ('official') ventures, and between those that are regulation-induced and those that are 'spontaneous market' initiatives. Public regulation-induced activity arises because of international agreements, such as the Biodiversity and Climate Change Conventions.

Table 1.1 sets out the resulting schema. The examples listed are dealt with more fully below.

4.1 Regulation-induced markets

The first way in which markets are emerging is via the existence of regulations or anticipated regulations. In turn, these regulations are international and national but since implementation is always at the national level we can treat them together.

Table 1.1. A Schema for Global Environmental Markets.

	Regulation-Induced	Spontaneous Market
Public/Official Ventures	Example: government to government measures under JI provisions of the Rio treaties: Norway, Mexico, Poland, GEF	Example: government involvement in market ventures: Swiss Green Export aid; debt-for-nature swaps
Private Sector Ventures	Example: carbon offsets against carbon taxes and externality adders	Example: purchase of exotic capital - Merck and Costa Rica deal on biodiversity.

Government-government trades

The first JI agreement has already been agreed between Norway, Poland and Mexico, through the medium of the Global Environment Facility (GEF). Norway agrees to create additional financing (through the revenues from its own carbon tax) for GEF carbon-reducing projects in Mexico (energy efficient lighting) and Poland (converting from coal burning to natural gas).[6] The US Environmental Defense Fund is understood to be developing a reforestation project in Russia. The US Government announced the *Forest for the Future Initiative* (FFI) in January 1993 under which carbon offset agreements will be negotiated between the USA and several countries, including Mexico, Russia, Guatemala, Indonesia and Papua New Guinea. The aim is for the US Environmental Protection Agency to broker deals involving the private sector.

[6] Global Environment Facility, 1992.

Private sector trades

The EU Draft Directive on a carbon tax and other European legislation also provides an incentive to trade in this way, as does State regulation on pollution by electric utilities in the USA. While not strictly a private enterprise trade, in The Netherlands, the state electricity generating board (SEP) established a non-profit making enterprise in 1990: *FACE* (Forests Absorbing Carbon Dioxide Emissions). FACE aims to sequester an amount of CO_2 equivalent to that emitted by one 600 MW power station. This is estimated to require some 150,000 ha: 5,000 ha in The Netherlands, 20,000 ha in central Europe and 125,000 ha in tropical countries. At the end of 1993 the actual area had risen to 180,000 ha with the additional area in tropical countries (see Table 1.2).

In the USA case the offset deals are currently not directly linked to legislation, but several have occurred which are clearly a mix of anticipation of regulation and 'global good citizenship'.[7] These include the *New England Power Co.'s* investment in carbon sequestration in Sabah, Malaysia through the reduction of carbon waste from inefficient logging activities. The forest products enterprise is run by Innoprise. New England Power estimate that some 300,000 to 600,000 tons of carbon (C) will be offset at a cost of below \$2 tC. *Rain Forest Alliance* will assist in monitoring the project. New England Power regard the Innoprise project as the first of a series aimed at assisting with the corporation's plan to reduce CO_2 emissions by 45% by the year 2000. *PacifiCorp*, an electric utility in Oregon, is considering reforestation projects and urban tree planting programmes in the USA, and an international sequestration project.[8] Two pilot projects have been announced: (a) a rural reforestation project in Southern Oregon which funds planting subject to a constraint of no harvesting for 45-65 years, at an estimated cost of around \$5 tC; and (b) an urban tree programme in Salt Lake City, Utah, at a provisionally estimated cost of \$15-30 tC sequestered. *Tenaska Corporation* is considering sequestration projects in the Russian boreal forests. Ultimately, some 20,000 ha of forests may be created in the Saratov and Volgograd regions at a cost of \$1-2 tC. Russian partners in the venture include the Russian Forest Service, the Ministry of Ecology and others. Tenaska is also planning reforestation projects in Washington State to complement a project in Costa Rica (see Table 1.2).

While these investments are aimed at CO_2 reduction, sequestration clearly has the potential for generating joint benefits, *i.e.* for saving biodiversity as well through the recreation of habitats. Much depends here on the nature of the offset. If the aim is CO_2 fixation alone, there will be a temptation to invest in fast growing species which could be to the detriment of biodiversity. It is important therefore to extend the offset concept

---

7 Newcombe and de Lucia, 1993.

8 Dixon et al., 1993.

so that larger credits are given for investments which produce joint biodiversity - CO_2 reduction benefits.

The US Energy Policy Act of 1992 requires the Energy Information Administration to develop guidelines for the establishment of a database on GHG offsets, together with an offset 'bank'. The Keystone Center in the USA is also establishing an interchange of information with a number of electric utilities to explore the issues involved in establishing offset deals.

4.2 Global good citizenship

Several offset deals appear to have been undertaken quite independently of legislation or anticipation of regulation. *Applied Energy Services (AES)* of Virginia has also undertaken sequestration investments in Guatemala (agroforestry) and Paraguay and is in the process of setting up another project in the Amazon basin. The Guatemala project is designed to offset emissions from a 1800 MW coal-fired power plant being built in Uncasville, Connecticut. The intermediary for the project is the World Resources Institute and in Guatemala the implementing agency is *CARE*. The project involves tree planting by some 40,000 farm families, soil conservation techniques, biomass conservation through fire prevention measures, *etc*. Carbon sequestration is estimated to be 15.5 mln. tons of carbon. The $15 mln. cost includes $2 mln. contribution from AES; $1.2 mln. from the Government of Guatemala; $1.8 mln. from CARE, with the balance coming in-kind from US AID and the Peace Corps.

The motivations for involvement vary. AES's involvement relates to its concern not just to offset CO_2 emissions, but to achieve local development and environmental benefits the deal brings. Dixon et al. report the sequestration cost as $9 tC overall, but inspection of the data suggests it may be much less than this.[9] $9 tC would be expensive for carbon sequestration alone, but there are other benefits from the scheme, including local economic benefits. In the Paraguay deal, AES has advanced money to the (USA) *Nature Conservancy* for investment in some 57,000 ha of endangered tropical forest. The International Finance Corporation (IFC) agreed to sell the land for $2 mln., well below the market price of $5-7 mln. AES expects to sequester some 13 mln. tC at around $1.5 per tC to offset CO_2 emissions from the Barbers Point 180 MW coal-fired

[9] Dixon et al., (1993).

plant in Oahu, Hawaii. Local benefits include eco-tourism, scientific research, recreation, agroforestry and watershed protection. AES is also planning to offset

Table 1.2. Private Sector Carbon Offset Deals.

Company	Project	Other Participants	Mln. tC sequestered or reduced	Total Cost $ mln.	$ tC sequestered
AES	Agro-forestry Guatemala	US CARE Govt. of Guatemala	15-58 over 40 years	15	a) 0.5-2* b) 1-4 c) 9
AES	Nature reserve Paraguay	US Nature Conservancy, FMB	13 over 30 years	6	a) 0.2 b) 0.45 c) <1.5
AES	Secure land tenure, sustainable agriculture Bolivia, Peru, Ecuador	Other utilities giving consideration to deal	na	2	na
SEP	Reforestation:				
	Netherlands		0.9	20	b) 22.7
	Czech Rep.		3.1	30	b) 9.7
	Malaysia	Innoprise	6.3	15.7	b) 2.5
	Ecuador		9.7	17.3	b) 1.8
	Uganda		7.2	8.0	b) 1.1
	Indonesia		6.8	21.7	b) 3.2
Tenaska and others	Reforestation: Russia	EPA, Trexler, Min.of Ecology, Russian Forest Service, *etc.*	0.5 over 25 years	0.5?	a) na b) 1 c) 1-2
Tenaska	Forest conservation in C.Rica + reforestation in Washington State	Other utilities giving consideration	na	5 +	na
PacifiCorp	Forestry, Oregon	Trexler	0.06 pa	0.1 pa	a) na b) na c) 5
PacifiCorp	Urban trees, Utah	Trexler, TreeUtah	?	0.1 pa	a) na b) na c) 15-30
New England PC	Forestry, Malaysia	Rain Forest Alliance, COPEC	0.1-0.15	0.45	a) na b) 3-4.5 c) na
New England PC	Methane recovery in Appalachians	na	na	na	na
Wisconsin Electric Power Co; NIPSCO Industry; Edison Dev. Co.	Coal to gas conversion	Bynov Heating Plant, Decin, Czech Republic	12,800 tpa ?	1.5	b) 43

a) Assumes 10% discount rate applied to total cost to obtain an annuity which is then applied to carbon fixed per annum, assuming equal distribution of carbon sequestered over the time horizon indicated;
b) Assumes no discounting.
c) Cost per tC as reported in Dixon et al. (1993).
* Barrett (1993a) estimates the total cost of sequestration at $0.73 tC.

Source: adapted from Dixon et al. (1993), Trexler (1989, 1993), Bann (1993), and FACE (1994). CO_2 converted to C at 3.67:1. Dutch guilders converted to $ at *f*1.75 per $.

emissions from a third power plant with support for indigenous peoples in Peru, Ecuador and Bolivia to secure title to their lands and to develop sustainable extractive activities.

Table 1.2 summarizes the private sector carbon offset deals to date.

4.3 Buying down private risk

Newcombe and de Lucia have drawn attention to another potentially very large private trade which has global environmental benefits.[10] Investment by the private sector in the developing world is invariably constrained by risk factors such as exchange rate risks, repayment risks, political risks and so on. In so far as this investment benefits the global environment, as with, say, the development of natural gas to displace coal, the existence of the risks reduces the flow of investment and hence the global environmental benefits. But these risks might be shared ('bought down') by having an international agency, such as the GEF, provide some funds or services which help reduce the risk. Given the scale of private investment flows, the potential here is enormous. Nor is there any reason why it should not benefit biodiversity, either indirectly as a joint benefit of other investments in *e.g.* raising agricultural productivity and hence in reducing the pressure for land degradation, or directly through afforestation schemes.

5 Problems with Joint Implementation

If the cost conditions are right, then JI promises significant cost savings in the context of climate change control. Why then does it cause controversy?

The problems appear to fall into several categories. The following list is far from comprehensive:[11]

- whether there is in fact much scope for trade: this is the issue of (i) whether costs of control vary significantly between trading Parties and (ii) transactions costs;
- the relationship between JI and developed country obligations under the FCCC;
- who should trade;
- the determination of cost;
- sinks vs emission reductions.

---

[10] Newcombe and De Lucia, 1993.

[11] For other questions see Jones, 1993.

5.1 The scope for trade

Various estimates exist of the scale of cost savings under a full emissions trading scheme.

In the USA, for example, it is known that abatement costs per ton of SO_2 vary substantially across states. Rico suggests that in the year 2000 a traditional command-and-control programme applied to the USA target of 10 mln. tons pa SO_2 reduction would cost some $3.2-4.9 bln. pa compared to costs with allowance trading of $1.1-2.1 bln., a saving of perhaps 60%.[12]

For sulphur trading schemes in the UN ECE region model simulations suggest that cost savings could be of the order of 2-5% of total costs without any trading and if restrictions are placed on depositions in third party countries, a problem that does not arise in the GHG context. The cost savings may be very much greater, up to 24%, if the restrictions on third party effects are less. This is, for example, the case if the sulphur emissions increases in the donor country and sulphur emissions reductions in the host country are equal, *i.e.* the 'exchange rates' for comparing reductions in different locations are in unity (see also section 1 of this chapter).[13]

In both cases, of course, significant trading is assumed, whereas most people's expectations appear to be that JI schemes will at least start modestly with a limited number of trades.

In the climate change context, RISO suggests marked variations in CO_2 control costs between European countries.[14] However, the variation is not in the direction that might be thought, namely lower costs in generally poorer countries. Abatement costs turn out to be lower, for example, in Germany, UK and Denmark than in Greece, Portugal and Spain. Barbier et al. found substantial variations in costs of CO_2 control between countries.[15]

Whether the evidence supports the assumption of marked cost differentials between countries or not is not a serious blow to JI as an 'enabling' concept. Essentially, all the agreements cited permit JI if Participants find it profitable to engage in it. It is not mandated in any sense. Hence if profitable opportunities exist they are likely to be exploited provided countries are aware of actual cost differentials. On the other hand, if cost differentials are small, or are 'perverse' in the sense that less developed countries

---

[12] Rico, 1994.

[13] UN ECE Task Force on Economics, 1994.

[14] RISO, 1992.

[15] Barbier et al., 1991.

have higher costs than developed countries, then JI does not hold out the prospect for major cost savings.

Even if cost differentials are significant, the potential for profitable trade is limited by transaction costs. Information on transactions costs is also difficult to evaluate. As noted above, the 'thinner' the market the higher are transaction costs likely to be. This suggests that there is a 'chicken and egg' problem in that the true cost saving benefits of JI will not be realized until large scale trading occurs, while large scale trading cannot occur until it has been demonstrated that JI results in significant cost savings.

5.2 JI and national obligations

It is easy to see that, from some standpoints, JI looks like an attempt by the developed countries to 'offload' their problem onto developing countries. This is the issue of whether or not JI should result in 'credits' for the donor country which are then set against its national target under the FCCC. If the credits are allowed, then the developed countries save costs by not reducing their own emissions as much as they otherwise would have done (see Figure 1.1). Low cost abatement options are then taken up in the developing countries.

Objections to such credits being counted appear to range from the purely practical to the moral. At the practical end of the spectrum is the concern that developing country low cost options would be used up so that developing countries would face only high cost options in the event that they signed up to their own emission targets at a future date. At the moral end of the spectrum is the view that global warming is the result of developed country profligacy, and that JI would postpone the radical changes in 'consumption' needed in the developed world.

A purely legalistic view correctly interprets the Convention as allowing JI against current emission reduction commitments. Equally, if JI is not to count against existing commitments, its likely scope will be fairly restricted since few countries have an incentive to go beyond their commitments. It is easy to overlook the fact that even with JI it is the donor country that is paying for the emissions reduction. Similarly, if JI brings benefits to the *host* country, those benefits will be forgone if JI is interpreted as being additional to national commitments. For these reasons, arguments that JI can be introduced on a voluntary basis without credits being counted against current commitments would seem injurious to the prospects for JI.

5.3 Who trades?

The common perception is that JI under the FCCC is about trade between developed countries and developing countries, and trade between developed countries and the economies in transition (Annex I countries in the FCCC). The obvious problem with this view, which would exclude inter-developed country trade, is that the developing countries do not have targets under the Convention. As such, it is not possible to say what they *would have done* if the JI deal did not take place. But determining what they would have done, the so-called 'baseline', is important since a credit for emissions reduction has no meaning unless it can be demonstrated that a genuine reduction has taken place. It is necessary to know that the emissions reduction under the JI deal is 'genuine', *i.e.* does not involve emissions reductions that would have taken place anyway.

This argument has been used by Klaassen to declare that non-signatories to the Second Sulphur Protocol should not be permitted to engage in JI deals.[16] Including non-signatories results in delays and high transaction costs. This appears also to be the experience with the Montreal Protocol[17] and is underlined by the fact that only one in ten proposed trades under the past US sulphur trading system was accepted in the form originally proposed. In other words, anything that opens the way for difficulties with authorization should be avoided or minimized.

Determining baselines will be a critical problem. Because of the problems of defining the counterfactual baseline, Bohm[18] argues that JI has the potential for being genuinely cost-reducing if the deals are between FCCC Annex I countries (developed plus transitional economies) but not if the deals are between developed and developing countries. This is because Annex I countries have (fairly) well defined emissions, base years and emissions targets, whereas this is not true for developing countries. All Parties to the FCCC are obliged to publish national programmes showing climate change mitigation measures and these could, in principle, be the means for determining the baseline in developing countries. There remains the problem of knowing if national programme statements themselves contain a degree of manipulation of true baselines.

Notice that the 'moral' objection to JI noted above largely disappears if trades are between Annex I countries only.

---

16 Klaassen, 1994b.

17 Klaassen, 1994b.

18 Bohm, 1994.

Offsetting this view is the fact that including non-signatories will enhance the comprehensiveness of any agreement and thereby maximize cost-savings. How far this is a serious concern in the FCCC context, however, is open to question since the vast majority of GHG emissions emanate from signatory countries.[19]

The other issue with who trades is whether trades occur between nations or companies, or both. The answer seems fairly clear that companies should be involved, and table 1.2 above has shown that this is already the case outside the terms of the FCCC.

5.4 The determination of cost

What cost do donors pay? The theoretical answer is that they pay the *incremental cost* incurred by the host country in implementing the JI deal. Incremental cost is always the cost of doing something that the host country would not otherwise have done. What it otherwise would have done produces the *baseline cost.* Unfortunately, whilst conceptually simple, incremental cost (Ic) is difficult to evaluate in practice. Both the Rio Conventions and the Montreal Protocol speak of 'agreed full incremental cost', but none of them defines the concept. It may be, however, that such definitions, while needed for the Convention generally, are not needed for JI.

The problem of the baseline has already been alluded to. It is only readily determined if countries have pre-established national plans. Jones suggests that no attempt should be made to determine the baseline or even a common methodology on costs.[20] But if so, it is difficult to see how it can be determined whether or not the JI deal has actually honoured the overall environmental objective, namely that GHG emissions should not increase. A JI deal might otherwise simply sanction a reduction that would have taken place anyway. This is particularly relevant for deals between Annex I countries involving economies in transition where structural change is likely to reduce GHG emissions in the baseline anyway. Some idea of the baseline therefore seems essential.

Incremental cost may also not define the actual transfer between donor and host, despite the implication of the FCCC that what is transferred, as funds or technology, is the 'full' incremental cost. This is because JI may involve the host country in securing some incremental domestic benefits that would not be present in its baseline. There are three options. First, any incremental domestic benefits (IDB) could be deducted from incremental cost so that the transfer (T) is given by T = Ic - IDB. Second, IDB could be ignored, in which case the host country secures a net benefit at least equal to IDB (at least, because the host country also gains its share of benefits from the reduction in

---

[19] Jones, 1993.

[20] Jones, 1993.

global warming), and T = Ic. Third, some fraction of IDB could be deducted, as is perhaps implied by the term 'agreed' full incremental costs, so that $T = Ic - \alpha IDB$.[21] The FCCC provides no guidance on this issue of benefit deductibility. Clearly, if IDB is deducted fully, then the host country has no real incentive to adopt a JI deal: it is neither better off nor worse off with the deal. On the other hand, most deals involving technology switching will bring updated capital equipment into being and the benefits of such investments are unlikely to qualify as IDBs.

However, whereas incremental cost needs to be addressed in the context of the FCCC as a whole, it is far from clear that operational guidelines are needed on it for JI deals. This is because JI is a fairly explicit bargain between host and donor.[22] They will therefore 'agree a price' at which the deal will take place. Issues such as incremental domestic benefits may well enter into those bargains, but there is no real need for an explicit set of guidelines on how to estimate incremental cost in this context. The transfer, T, will simply be agreed between parties, each of whom knows whether or not it is to their advantage.

5.5 Sinks vs emissions reductions

In principle, a JI deal could involve augmenting a sink or securing an emissions reduction. The existing deals listed in table 1.2 are generally of the kind where emissions are traded for afforestation, reforestation or efficient forest use. Bohm argues that the uncertainty surrounding the true measure of carbon fixed will simply add to the monitoring costs of JI and hence make JI less attractive.[23] As such, he favours omitting them from JI deals under the FCCC at least for now.

Against this, one has to ask why it is that the private deals have opted for this option. A cynical view might be that such deals are capable of manipulation by all parties in their own interests. Certainly, JI carries with it the risk of threat situations when the baseline is not readily identifiable. Thus a host country might refuse to afforest and thus claim that 'no afforestation' defines its baseline. It can then attract an afforestation JI project and claim that newly planted trees are a genuine contribution to carbon fixation.

The situation might be worse still if JI is extended to 'avoided deforestation', with threats to deforest being rewarded by JI deals. The donor agent also has an incentive to exaggerate the extent of host country emissions reduction or fixation since it gains more

---

21 Pearce and Barrett, 1994.

22 Bohm, 1994a.

23 Bohm, 1994a.

by way of its own emissions increases that will be allowed. But actual deals to date seem to have been motivated more by a concern to operate at a manageable level, to demonstrate JI in a context where the donor does something different to its normal operations (effectively making the baseline more credible), and to generate ancillary development benefits. Forestry projects appear to meet these conditions more readily than, say, fuel switching investments.

6 Conclusions

Clearly, JI is far from being straightforward in its application. The particular problems that arise may be summarized as those of designing compatible incentives for the revelation of 'genuine' trades; monitoring; determining who trades and what the gains from trade are; and providing assurances to the developing world that JI is not a mechanism whereby the rich world can escape its own obligations. The issue is whether these problems are worth solving and that can only be resolved by securing a clearer idea of what the gains are to each party: the cost savings to the donor and the world at large, and the developmental gains to the host countries. We might end on the general reflection that if we ignore the cost savings argument, or if we allow too many restrictions to be placed on JI deals, we may risk reducing the incentives of the developed country Parties to commit themselves seriously to strengthening the existing Conventions, and to any new ones that may have larger benefits to developing countries. As with all things, compromise has to be the outcome.

References

Bann, C., 1993. *The Private Sector and Global Warming Mitigation*, Centre for Social and Economic Research on the Global Environment, University College London, mimeo.

Barrett, S., 1993a. *Joint Implementation for Achieving National Abatement Commitments in the Framework Convention on Climate Change*, Report to the Environment Directorate, Organization for Economic Cooperation and Development (OECD), Paris; London Business School and Centre for Social and Economic Research on the Global Environment, University College London, mimeo.

Barrett, S., 1993b. *A Strategic Analysis of 'Joint Implementation' Mechanisms in the Framework Convention on Climate Change*, Report to the United Nations Conference on Trade and Development (UNCTAD), Geneva. London Business School and Centre for Social and Economic Research on the Global Environment, University College London, mimeo.

Barbier, E., J. Burgess and D.W. Pearce, 1991. Technological Substitution Options for Controlling Greenhouse Gas Emissions, in R.Dornbusch and J.Poterba, *Global Warming: Economic Policy Responses*, MIT Press, London and Cambridge, 109-160.

Bohm, P., 1994a. *On the Feasibility of Joint Implementation of Carbon Emissions Reductions*, Department of Economics, University of Stockholm, *mimeo.*

Bohm, P., 1994b. *Making Carbon Emissions Quota Agreements More Efficient: Joint Implementation vs Quota Tradeability*, Department of Economics, Stockholm University, *mimeo.*

Dinerstein, E. and E. Wickramanayake, 1993. Beyond 'Hotspots': How to Prioritize Investments to Conserve Biodiversity in the Indo-Pacific Region?, *Conservation Biology*, Vol.7, No.1, March, 53-65.

Dixon, R., K. Andrasko, F. Sussman, M. Trexler and T. Vinson, 1993. Forest Sector Carbon Offset Projects: Near-Term Opportunities to Mitigate Greenhouse Gas Emissions, *Water, Air and Soil Pollution.*

FACE (Forests Absorbing Carbon Dioxide Emission), 1994. *Annual Report 1993*, Arnhem.

Global Environment Facility, 1992. *Memorandum of Understanding on Norwegian Funding of Pilot Demonstration Projects for Joint Implementation Arrangements Under the Climate Convention*, GEF, World Bank, Washington DC, mimeo.

Jones, T., 1993. *Operational Criteria for Joint Implementation*, OECD, Paris, June.

Klaassen, G., 1994a. Trade Offs in Sulfur Emission Trading in Europe, *Environmental and Resource Economics*, Special Issue on Acid Rain (edited by G. Klaassen and D.W. Pearce), forthcoming.

Klaassen, G., 1994b. *Joint Implementation in the Second Sulfur Protocol: a Tempest in a Teapot?*, International Institute for Applied Systems Analysis (IIASA), Laxenburg, Austria.

Newcombe, K. and R. de Lucia, 1993. *Mobilising Private Capital Against Global Warming: a Business Concept and Policy Issues*, Global Environment Facility, Washington DC, mimeo.

Pearce, D.W. and S. Barrett, 1994. *Incremental Cost for Biodiversity Conservation*, Global Environment Facility Working Papers, forthcoming.

Rico, R., 1994. The US Allowance Trading System for Sulfur Dioxide: an Update on Market Experience, *Environmental and Resource Economics*, Special Issue on Acid Rain (edited by G. Klaassen and D.W. Pearce), forthcoming.

RISO, 1992. *UNEP Greenhouse Gas Abatement Costing Studies*, RISO National Laboratory, Denmark, August.

Sorrell, S., 1994. *Pollution in the Market: the Theory and Practice of Emissions Trading for the Control of Air Pollution*, Science Policy Research Unit, Brighton UK.

Trexler, M., P. Faeth and J.M. Kraemer, 1989. *Forestry as a Response to Global Warming*, World Resources Institute, Washington DC.

Trexler and Associates, 1993. *Forestry and Carbon Offset Strategies*, Report prepared for Environmental Affairs Division, Edison Electric Institute, Washington DC.

UN ECE, 1994. *Economic Aspects of Abatement Strategies: Progress Report by the Chairman of the Task Force*, UN ECE, Geneva.

2. The Scope for Joint Implementation in 'Energy for Tomorrow's World'

MICHAEL JEFFERSON*

1 Introduction[1]

Two WEC publications form the basis for this assessment: the WEC Commission report 'Energy for Tomorrow's World' (ETW), first published in September, 1993; and the WEC book 'New Renewable Energy Resources', published in July, 1994. The WEC Commission report was based on the work of over 500 people from all over the world. It was both top-down and, more importantly and uniquely, bottom-up: nine Regional Working Groups, covering the globe, produced reports as input to the Commission's work. These regional groups and their work reflected local experience, realities and priorities.

An interim report from the WEC Commission was debated by over 4,500 delegates from 130 countries at the 15th WEC Congress in Madrid, Spain in 1992. As a result a new case illustrating possible future global energy demand and supply prospects out to 2020 was developed, and other issues brought to the fore.[2]

The WEC report on new forms of renewable energy (which *excludes* traditional fuels such as fuel wood, crop wastes and animal dung, together with large hydro over 10 MW, but *includes* solar, modern biomass, wind, geothermal, ocean/tidal and small hydro below 10 MW) was also a global effort, involving more than 80 technical specialists. Again, this work was conducted within a framework of nine regions covering the world.

---

* Michael Jefferson is Deputy Secretary-General at World Energy Council, London, UK.

1 The views expressed here are based upon work done by the World Energy Council (WEC). The WEC is a Non-Governmental Organisation, non-profit making, and multi-energy in membership and objectives. The broad mission of the WEC is the promotion of the sustainable supply and use of energy for the greatest benefit of all. Based in London since its creation in 1924, the WEC is global in coverage, with Member Committees in about 100 countries. Some 55% of the membership is comprised of currently rapidly industrialising and developing countries, and 16% from the economies in transition of the former Soviet Union, Central and Eastern Europe. The views expressed here are, therefore, not a narrow OECD perspective but seek to balance a broader range of perceived priorities by adoption of a wider analysis and agenda for Joint Implementation (JI).

2 The WEC is currently reviewing its longer-term assessment jointly with the International Institute for Applied Systems Analysis (IIASA), Laxenburg, Austria. The nine world regions have now been extended to 14, to capture some of the finer details and contrasts within regions.

C. J. Jepma (ed.), The Feasibility of Joint Implementation, 33–55.

These introductory comments are offered because they help to explain the nature of the process we have gone through and the conclusions arrived at. In particular they explain a balance of priorities which may seem at odds with the prime thrust of the Intergovernmental Panel on Climate Change (IPCC) and the UN Framework Convention on Climate Change (FCCC). In fact the WEC has adopted an objective stance towards the issue of potential climate change, a position akin to IPCC Working Group I on the Scientific Assessment - notably as set out in the updated assessment of 1992 and, in more recent work, on the radiative forcing of climate now in draft for review. Similarly, there is nothing in what follows which runs fundamentally counter to the principles of the FCCC as set out in Article 3. In particular, there is a demand for change, and for action now, which is fully compatible with the principle that lack of full scientific certainty should not be used as a reason for postponing relevant measures.

The differences lie in the clear awareness that for most people in the world, and for most countries, potential climate change is not currently a high priority. There are other, more pressing and therefore higher, priorities. In the energy field a variety of actions - to raise efficiency in supply and use, to introduce cleaner conversion processes and reduce local and regional pollutants, and to accelerate the provision of renewable forms of energy - will both meet these other priorities and assist with mitigation of potential climate change.

This contribution therefore concludes that Joint Implementation projects which have as their first and principal aim the mitigation of or adaptation to climate change, have an aim which does not accord with the priorities of most people and countries. If climate change uncertainties continue beyond a decade, or become generally regarded as of little consequence (however unlikely that may now appear), then Joint Implementation schemes justifiable on other grounds will lose their main rationale and impetus. However, if potential climate change is employed as a concern which backs up other reasons for action, which are commonly accorded high priority and urgency, then the reasoning becomes more robust and sustained. What is required, therefore, is a broader agenda for Joint Implementation schemes which still conforms to the main principles of the FCCC.

In what follows, the wider context and priorities of that broader agenda will emerge. Some may feel that the case for JI within the climate change context has become transformed into a renewed case for Official Development Assistance (ODA) focused on energy-related schemes. It is widely considered that there has been too little ODA in the past three decades, and much of what has taken place has been counter-productive. Nevertheless, incremental investments in countries least able to attract finance and technology, through their own lack of resources and impoverishment, may well have merit - not least where energy efficiency gains are made, cleaner conversion technologies are put in place, renewable energy provision is expanded, energy availability is increased, and the metering and pricing of energy use are improved. JI schemes which mitigate the worst examples of energy-related pollution elsewhere may also have merit. And an approach which seeks to address the various priorities and goals is likely to attract more support than one which has a priority and main goal that is not generally shared. How far such schemes should be entirely commercial, and what the scope is for transfers of finance and technology, is another large subject not covered here.

2 Global Energy Demand

People need energy for the services it provides - heating, cooling, cooking, mobility, lighting and motive power. More than half the world's current population of 5.5 bln. does not have access to the commercial energy which provides these services and satisfies basic needs. World population is expected to increase, and material expectations and desires are expected to add to the upward driving forces on demand.

Population, energy per capita availability and use, and lifestyle shifts are just some of the issues which create uncertainty about the future evolution of global energy demand. Finance, technology, the whole range of environmental impacts and concerns, and institutional changes are among the others. Let us look at just two of these: world population projections and energy efficiency prospects.

The UN makes world population projections (Figure 2.1). In the UN High Projection world population reaches 17.6 bln. by the year 2100, and over 25 bln. by 2150. It is worth noting that this is the population assumption on which the IPCC's IS92f scenario is based. In the UN Low Projection world population is as little as 6.4 bln. by 2100, and below 5 bln. (thus less than the current figure) by 2150. It is this Low Projection upon which the IPCC's low CO_2 emissions and low atmospheric CO_2 concentrations scenarios IS92c and IS92d are based.

The WEC Commission took broadly the UN Medium Projection (which is much the same as the World Bank's 'Base Case') as the common population assumption for all its cases. Thus world population is assumed to grow to about 8 bln. in 2020, 8.5 bln. in 2025, 10 bln. in 2050, and approaching 12 bln. in 2100. This population assumption is reasonably close to the IPCC's scenarios IS92a, b and c (with world population reaching 11.3 bln. in 2100).

These widely different population assumptions underpinning the IPCC's current emissions and atmospheric concentrations of greenhouse gases (GHG) are often overlooked, yet they are clearly of critical importance. As will be seen, the WEC Commission's Ecologically Driven Case C is much more challenging, and has much lower emissions and concentrations, than any IPCC Scenario on compatible assumptions.

The WEC Commission recognized that it could not forecast the future. Instead, four cases were developed to 2020: High Growth (A), Modified Reference (B_1), Reference (B), and Ecologically Driven (C). The basic assumptions are set out in table 2.1.

To limit the length of this contribution, discussion is confined to the energy intensity reductions assumed. By this is meant the ratio of energy consumption to Gross Domestic Product expressed in constant prices. This is a very rough measure of energy efficiency. There are all sorts of questions which can arise about national income and expenditure accounting; whether measurement at market prices and exchange rates is more or less relevant than a purchasing power parity basis (the latter is clearly more meaningful for many developing countries); how far current measurement fails to capture key features - for instance, externalities such as environmental impacts; and how far measurement may change in future to capture such features and thereby substantially change the definition and measurement.

Figure 2.1. UN World Population Projections.

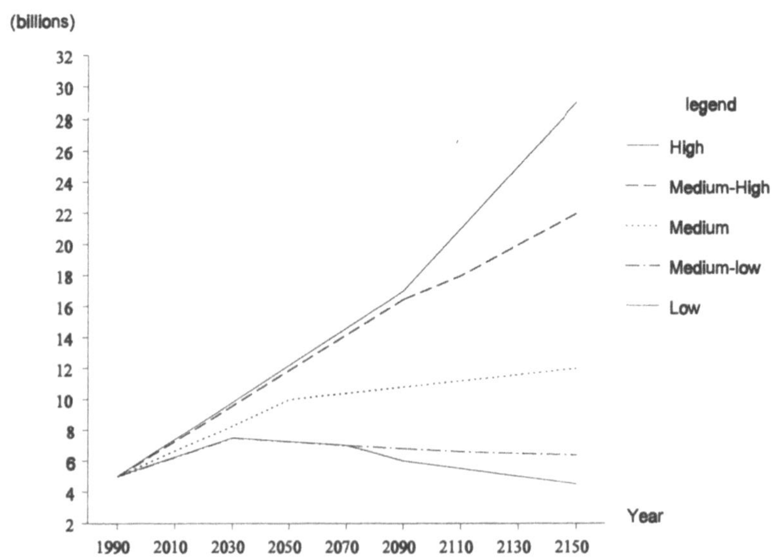

Table 2.1. WEC Commission's 4 cases.

Case	A	B	C	D
Name	High Growth	Modified Reference	Reference	Ecologically Driven
Economic Growth % pa	High	Moderate	Moderate	Moderate
OECD	2.4	2.4	2.4	2.4
CEE/CIS	2.4	2.4	2.4	2.4
DCs	5.6	4.6	4.6	4.6
World	3.8	3.3	3.3	3.3
Growth per Capita				
OECD	Moderate	Moderate	Moderate	Moderate
CEE/CIS	Moderate	Moderate	Moderate	Moderate
DCs				
Asia	Very High	High	High	High
Sub-Sahar. Africa	Moderate	Low	Low	Low
Most Others	High	Moderate	Moderate	Moderate
Energy Intensity Reductions % pa	High	Moderate	High	Very High
OECD	1.8	1.9	1.9	2.8
CEE/CIS	1.7	1.2	2.1	2.7
DCs	1.3	0.8	1.7	2.1
World	**1.6**	**1.3**	**1.9**	**2.4**
Technology Transfer	High	Moderate	High	Very High
Energy Efficiency Improvement				
OECD	High	High	High	Very High
CEE/CIS	Moderate	Moderate	High	Very High
DCs	Moderate	Moderate	High	Very High
Institutional Improvements (World)	High	Moderate	High	Very High
Possible Total Demand (Gtoe) (1990 = 8.7 Gtoe)	Very High 17.2	High 16.0	Moderate 13.4	Low 11.3
CO_2 Emissions from Fossil Fuel (GtC) (1990 = 5.9 GtC)	11.5	10.2	8.4	6.3

It is clear that economic growth over time reflects changes in the structure of economies. Historically, energy intensity reductions have principally reflected shifts in economic structure (from agricultural to manufacturing, then to service economies for the leading OECD member countries), and 'normal' capital-stock turnover not greatly influenced by considerations of energy efficiency or energy prices.

The WEC Commission did not assume for any of its cases that Business As Usual[3] was a serious and acceptable outlook. Energy intensity reductions between double and treble the historic rate for the OECD countries were assumed.

These are very challenging assumptions if we consider the recent past. Since 1986 there has been either an increase in energy intensity, or little change, in some regions of the world: the Middle East and North Africa, Latin America and Sub-Saharan Africa. This is despite the fact that in over 50 countries in the world over the past 20 years energy per capita use has actually declined, for reasons which have nothing to do with energy efficiency per se.

In several OECD countries overall energy intensity has also increased since 1988, principally due to recession but also a result of low energy prices. The UK, France and The Netherlands are in this category. Even in Japan, manufacturing industry's energy intensity has risen since 1989.

Against this discouraging background it cannot reasonably be argued that efficiency and pollution mitigation measures will *either* take place in OECD countries *or* JI schemes will be implemented in developing countries and economies in transition. It must be *both*, and on a scale and at a rate which have never previously been achieved.

There is, after all, enormous scope for energy efficiency improvements. Various researchers have concluded that about 95% of energy produced is wasted or used unproductively. About 80% of the energy used in a typical passenger car on a typical range of journeys is unproductive - lost in friction and thermodynamic inefficiency. In manufacturing industry 75% to 90% of energy is typically unproductive or wasted. Light bulbs have attracted considerable interest in this workshop: the standard incandescent light bulb has an efficiency of 3% if used efficiently, and less than 1% if not used efficiently. In coal-fired power generation the global conversion efficiency is about 32%, and in the OECD area about 36%, but only 20-22% in some developing countries and economies in transition.

Whereas overall energy intensity tends to decline over time, reflecting increased national income and structural change, the energy intensity of the transport sector has tended to rise virtually everywhere over the past two decades. In the EU overall energy intensity has declined by some 1.6% per annum since 1974, but transport's energy intensity has risen 0.8% per annum. If the transport sector had been excluded, the EU would have experienced energy intensity reductions of 2.6% per annum since

---

[3] The 'Business as Usual' scenario has been produced by the Response Strategies Working Group of the First Assessment of the IPCC in 1990. According to the scenario, which assumes that few steps are taken to limit emissions, equivalent CO_2 in the atmosphere will have doubled by around the year 2025.

1974. One needs only to recall that there are 600 people for every car in China, 120 people for every car in India, 16 people for every car in the CIS Republics, to realize how critical the future of transportation is for sustainable development. (By comparison there are 2 or 3 people to every passenger car in Western Europe, and 1.7 people to every car or small truck in the USA.)

Given that there have been some misleading figures given at this workshop on historic changes in energy intensity, it should be noted that whereas overall energy intensity in the United States (including wood burning) has been in decline since 1850 at latest, this reduction stagnated in the period 1895-1920. This stagnation, combined with the fact that some early research in this field covered US manufacturing industry only, has led to the idea that (a) US energy intensity only began to fall after 1920; and (b) the decline has been steady. A similar story emerges from the UK, where energy intensity began to decline by 1870 but actually rose in the period 1914-1920.

These experiences lead to a questioning of the widespread view that energy intensity in developing countries continues to rise in general. Once traditional biomass energy is included, it seems clear that over the past 25 to 40 years many developing countries have posted energy intensity declines: India, Thailand, Philippines, and - more recently - China, for example.

Again, these declines should not occasion too much surprise if the assessments of the World Bank and others are accepted: that 20% to 30% gains in energy efficiency are feasible in most developing countries simply by using existing plant and equipment more efficiently. Nevertheless, there remain huge imbalances in energy per capita availability and use around the world. The South Asian region uses one-twentieth the energy per capita of North America. Individual differences within the regions are far greater.

The WEC Commission was rather conservative, in the eyes of some critics, about energy demand developments in developing countries. This conservatism was based partly upon concerns about the availability of finance and technology. The resulting energy per capita implications are set out in Table 2.2. The low figure of 0.92 toe (ton oil equivalent) per capita figure for the developing countries in the Ecologically Driven Case C by 2020, apparently hardly higher than the 1990 figure, should be set against a background of a 75% assumed increase in energy efficiency as well as the fact that over 90% of the population growth to 2020 is expected to occur in the developing countries.

Table 2.2. Energy per Capita (toe per capita).

	1990	2020			
		A	B_1	B	C
OECD	4.73	4.82	4.82	4.82	3.72
DCs	0.73	1.33	1.16	1.03	0.92
World	1.66	2.13	1.98	1.65	1.39

As a result of considering the various issues, and applying the various assumptions, the WEC Commission concluded that at least 85% of the increase in global energy demand 1990-2020 would occur in the developing countries. Up from 8.8 Gtoe (gigatons oil equivalent) in 1990 to a range of 11.3 Gtoe to 17.2 Gtoe in 2020. The figures are set out in figure 2.2.

Figure 2.2 Energy demand until 2020.

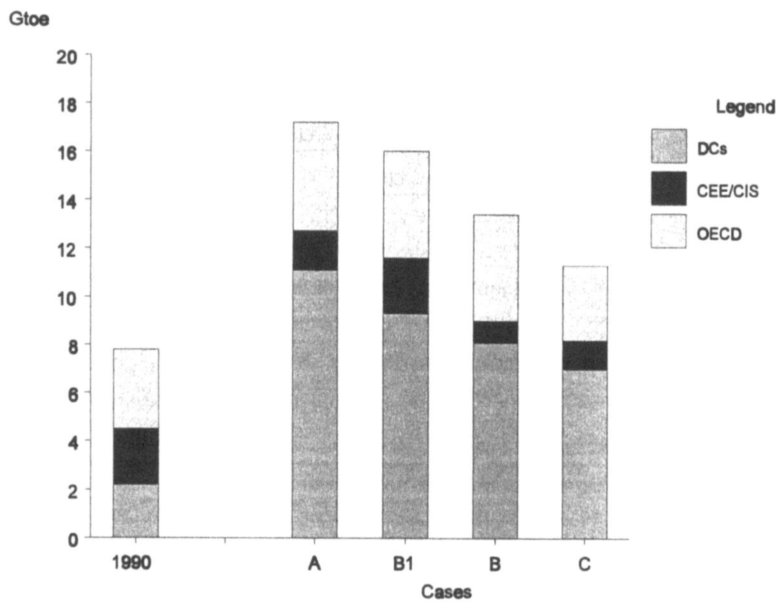

3 Global Energy Supply

The WEC Commission naturally gave careful consideration to the supply options, taking into account: the heavy reliance on fossil fuels at present; the preferences of many countries rich in coal and oil resources; public concerns with safety, waste disposal and proliferation issues surrounding nuclear power generation; and the length of time it will take to expand new renewable energy provision to the point where it can make a really substantial contribution.

Table 2.3. Energy Mix - Past and Future Global Annual Fuel Use (Gtoe).

	1960	1990	in 2020			
			A	B_1	B	C
Coal	1.4	2.3	4.9	3.8	3.0	2.1
Oil	1.0	2.8	4.6	4.5	3.8	2.9
Natural Gas	0.4	1.7	3.6	3.6	3.0	2.5
Nuclear	-.-	0.4	1.0	1.0	0.8	0.7
Large Hydro	0.15	0.5	1.0	1.0	0.9	0.7
'Traditional'	0.5	0.9	1.3	1.3	1.3	1.1
'New' Renewables	-.-	0.2	0.8	0.8	0.6	1.3
Total	3.3	8.8	17.2	16.0	13.4	11.3

Table 2.3 compares the global fuel mix in 1960, 1990 and the four WEC Cases in 2020. The work on new renewables reflected the study 'New Renewable Energy Resources' by the WEC's Renewable Energy Resources Committee. New renewable forms of energy provided only 2% of global primary energy supply in 1990. Under a 'Current Policies' Scenario this figure by volume could increase nearly four-fold by 2020. Accelerated development, propelled by the desire to reduce dependency on fossil fuels more quickly, could lead to a nine-fold increase. These are described as the minimum and maximum contributions by 2020 in table 2.4.

Table 2.4. Minimum/Maximum Contributions from 'New' Renewable Energy[a].

	In 2020 'Minimum'		In 2020 'Maximum' with major policy support	
	Mtoe	% of total	Mtoe	% of total
'Modern' Biomass	243	45	561	42
Solar	109	20	355	26
Wind	85	15	215	16
Geothermal	40	7	91	7
'Small' Hydro	48	9	69	5
Oceanic	14	3	54	4
Total	539	100	1345	100
% of Total Energy Demand	3-4		8-12	

[a] In 1990 'New' Renewables contributed 164 Mtoe (1.9%) of total energy demand.

It is not only the fact that new renewable energy provision starts from a low base and will take time to reach a substantial volume. There are severe local environmental impacts associated with most forms of renewable energy, if not all. Environmental awareness and constraints are likely to be greater in the future than ever in the past, for all forms of energy provision. Renewable forms of energy are likely to be affected by such awareness and constraints, and therefore great efforts will be required to avoid the more severe impacts. Renewable energy provision is likely to fall well below its technical potential for this reason. Nevertheless, unit costs are expected to come down substantially; and dispersed, local and relatively cheap renewable energy investments are likely to be very attractive to many developing countries - especially those with a comparative advantage in solar insolation, biomass growth, or hydro (subject to the environmental and people displacement problems caused by large hydro schemes).

Contrary to the impression given by Greenpeace International, in Chapter 5 of this volume, it is improbable that Joint Implementation schemes will be directed to nuclear and large hydro schemes. This is: because developing countries are likely to be where the most rapid development of renewable energy development takes place; because those countries will be anxious to avoid very large single schemes which result in substantial external indebtedness; and because JI is more likely to be directed to a large number of smaller schemes in the more impoverished developing countries.

4 Long-Term Global Energy Prospects

Figure 2.3. Global Primary Energy Demand (Gtoe) to 2100: WEC Commission Cases A, B and C.

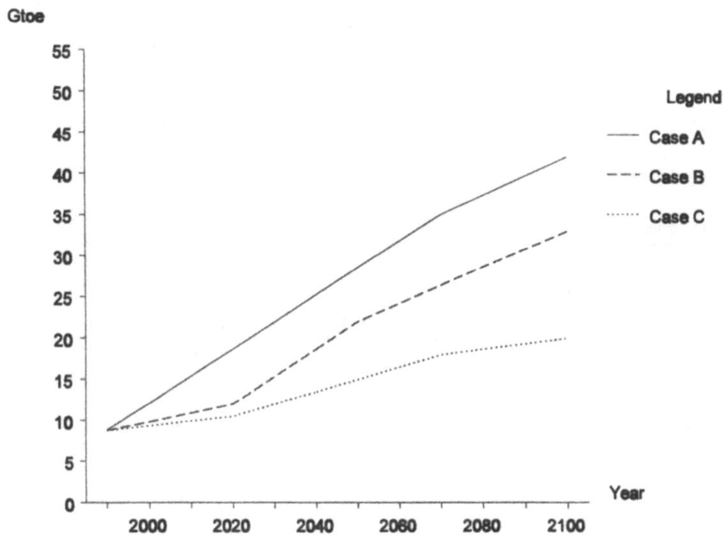

Because a number of energy-related issues either are unlikely to become of chronic concern before 2020, though perceived as potentially threatening thereafter, or because effective and substantial responses are not realisable within such a short time-frame, the WEC Commission looked at demand, supply and other possibilities in the much longer term.

Figure 2.3 provides the WEC Commission's necessarily speculative views on global primary energy demand possibilities for the three cases taken out to 2100. There is clearly a wide range of possible outcomes. Table 2.5 gives the fuel mix assumptions associated with those demand profiles. It will be noted that in the Ecologically Driven Case C the possibility of new renewable forms of energy accounting for as much as 50% of global primary energy supply by 2100 is recognized. Table 2.6 provides a possible breakdown by forms of new renewable energy, for selected years out to 2100. It will be out to 2100. It will be noted that over the next century the various forms of solar energy are expected to expand to become the major provider of renewable energy.

Table 2.5. Long-Term Global Energy Demand (and main energy sources).

		A		B		C	
	1990	2050	2100	2050	2100	2050	2100
Energy Demand (Gtoe)	8.8	27	42	23	33	15	20
Fossil Fuels (% of primary energy)	77	58	40	57	33	58	15
Nuclear (% of primary energy)	5	14	29	15	28	8	11
New Renewables (% of primary energy)	2	15	24	14	26	20	50

Table 2.6. Long-Term 'New' Renewable Energy Possibilities (Mtoe).

	2020	2050	2075	2100
		'Medium'		
Modern Biomass	243	440	950	1,700
Solar	109	620	1,650	3,500
Wind	85	350	850	1,400
Other	102	240	450	650
Total	539	1,650	3,900	7,250
		'High'		
Modern Biomass	561	1,000	1,600	2,400
Solar	355	1,535	3,200	5,000
Wind	215	600	1,100	1,800
Other	214	400	650	800
Total	1,345	3,535	6,550	10,000

This will, however, be a costly business. Even to the year 2020 substantial capital investments will be required. Table 2.7 sets out some estimates under the Current Policies and Ecologically Driven scenarios (Minimum and Maximum Contributions) mentioned earlier. If anything, the WEC Commission believed these estimates are likely to prove on the low side.

Table 2.7. Capital Requirements for New Renewables to 2020 (Cumulative $ Bln. (1990 Prices)).

	Steady Evolution			Major Policy Support		
	2000	2010	2020	2000	2010	2020
Solar	52	134	313	65	265	1205
Geothermal	15	20	35	20	60	110
Modern Biomass	50	100	150	66	140	260
Ocean	1	10	55	1	50	150
Small Hydro	21	50	100	36	88	150
Subtotal	159	374	833	223	738	2280
Transmission	10	23	56	15	49	147
Total	169	397	889	238	787	2427

As a major part of this expenditure is likely to be incurred by, or on behalf of, developing countries likely to find difficulty in attracting the necessary finance and technology, the scope for JI initiatives will be obvious. Beyond 2020 the scale of investments required will, of course, rise still further.

Figure 2.4. Annual CO_2 Emissions from Fossil Fuel Combustion (GtC) 1990-2100.

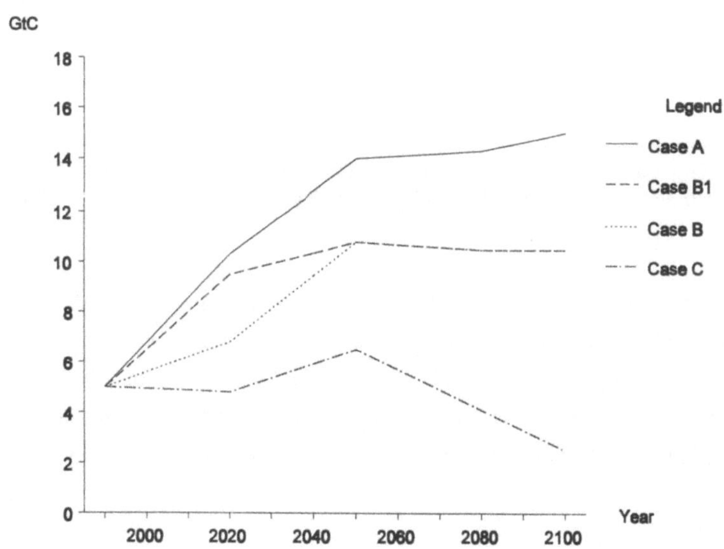

5 CO_2 Emissions and Concentrations

The WEC Commission then produced estimates of annual CO_2 emissions from fossil fuel combustion out to 2100 under its three long-term cases. In all three, global annual CO_2 emissions continue to rise until at least 2050. This is principally due to developments in the current rapidly industrialising and developing countries. Figure 2.4 provides the results.

Elsewhere in this volume, attention will be paid to CO_2 emissions from cement manufacture (part III). Table 2.8 provides the estimates for fossil fuel combustion, natural gas flaring and cement manufacture using WEC Commission data plus input by Gregg Marland of Oak Ridge National Laboratory, USA. Natural gas flaring and cement manufacture are felt to be marginal in this context.

Table 2.8. CO_2 Emissions from Fossil Fuel Combustion, Natural Gas and Cement Manufacture (GtC).

(1990 = 6.1)	Case			
In Year:	A	B1	B	C
1995	6.79	6.67	6.45	6.15
2000	7.57	7.29	6.83	6.21
2005	8.43	7.98	7.22	6.27
2010	9.40	8.74	7.65	6.34
2015	10.49	9.56	8.09	6.41
2020	11.70	10.47	8.57	6.49
2025	12.05	10.91	9.04	6.64
2050	15.06	15.06	12.39	7.48
2075	14.96	14.95	11.82	4.71
2100	16.50	16.50	11.68	2.57

Source: Energy Data - WEC; Carbon coefficients, NG flaring and cement production (2.58% of total) plus input by Gregg Marland, Oak Ridge National Laboratory.

So far as atmospheric CO_2 concentrations are concerned, using the MAGICC general circulation model of Wigley, the WEC Commission's Case A was found to be very close to the mid-point of the IPCC's Scenario IS92a (see Figure 2.5 in which the range of the IS92a scenario is indicated by the area between the upper and lower bounds IS92a). Case B was somewhat lower and case C was substantially lower.

Figure 2.5. Atmospheric CO_2 Concentrations (ppmv) to 2100.

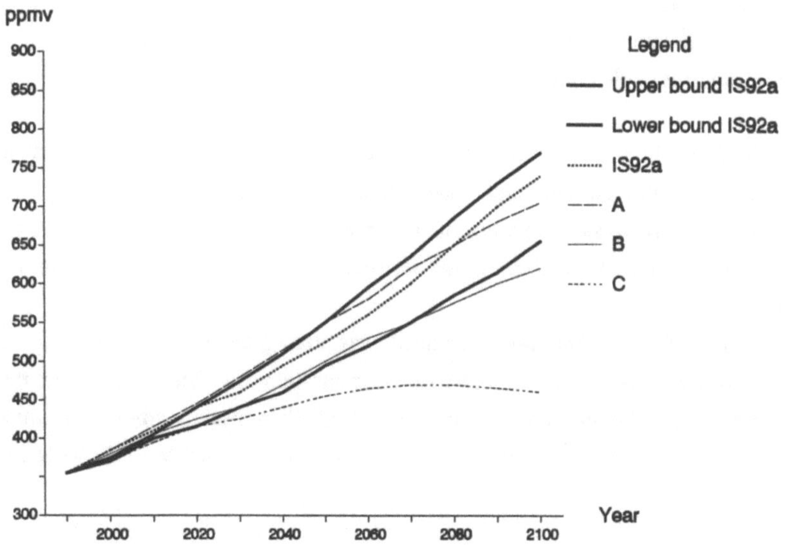

What is important here is not the demonstration that the world's energy industry is comfortable with examining the possibilities relating to potential climate change (though this may occasion surprise in some quarters), but what the shorter-term implications are. Between 1990 and 2020 on these calculations, under Case A, the rise in atmospheric CO_2 concentration is greater than has occurred over the past 200 years. Even under Case C the rise is two-thirds of the past 200 years.

These possibilities (the uncertainties being fully recognized) indicate that action must not be delayed. The WEC Commission was absolutely clear that action is needed *now*. It follows that a number of other contributors to this volume who will call for steps to be phased-in over 10 or 20 years, or delayed altogether, have a stance incompatible with that of the WEC Commission. It is recognized that the idea of a two-phase approach has attracted interest since the McKinsey Report to the Noordwijk Conference on Atmospheric Pollution and Climate Change, in November 1989.

But what action needs to be taken urgently? Here it is necessary to point out that for eight out of the WEC Commission's nine regions potential climate change was not a high priority (the exception was Western Europe). There were several, higher and more pressing priorities like food availability per capita - in decline in sub-Saharan Africa for some 25 years. Health provision and employment also ranked high. Then there were several local and regional pollution issues:

- uncollected solid waste (affecting some 2.2 bln. people);
- contaminated water (affecting over 1.1 bln.);
- lack of sanitation/sewage treatment (affecting over 1.7 bln.); and
- domestic fuel collection and use.[4]

Then there are the more familiar energy-related pollution problems of city smog, lead pollution, acid rain, and land and water pollution more generally.

These items have been at the top of the development and aid agenda for at least 30 years. They remain at the top. JI can be used as a means of reinvigorating some of them.

What the WEC Commission concluded was that general expectations of curbing global GHG emissions and atmospheric concentrations of these gases over the next few decades are largely unrealistic. The WEC Commission advocated immediate precautionary action on a 'minimum regret' basis, but this had the following implications:

- for as long as uncertainties remain about climate change, the benefits of action will remain uncertain; and
- costs will be borne in advance of benefits but, more importantly, if action is taken which is relevant to a range of problems (not just fears of climate change) then the costs of mitigating climate change cannot be accurately ascertained.

6 Key Requirements of JI

What is clearly required is a change across a broad front: of attitudes, systems, dimensions, energy development planning, and governments' role in order to encourage change. But this means that people - as energy consumers and voters - recognize and support the need for change. There are a number of concerns here:

- an insufficient sense of urgency;
- too narrow a perspective of problems;
- failure to recognize the plight of the South - especially of the most impoverished, and politically weakest, countries;

---

[4] Women in several sub-Saharan African countries regularly walk 4 to 8 hours per day, over distances up to 45 km, to collect a diminishing supply of fuel wood; some 6 million people die in India annually due to respiratory and related problems caused by burning traditional fuels).

- although JI projects may only prove of modest incremental benefit globally, they could provide the lifeline for the most deprived countries;
- JI projects could also be the means of identifying and tackling some of the world's most polluting plants; and
- broadening the agenda for JI projects implies an expansion of focused Overseas Development Assistance.

What are the prospects for requisite change and the obstacles thereto?

First, governments and intergovernmental institutions should review whether there is consistency between policies and goals for energy on the one hand, and the environment on the other. For instance, the 1994 edition of the International Energy Agency's Outlook states:[5]

> The rate of decline of energy intensities in the OECD is expected to be less than the average over the previous twenty years, as recent evidence indicates.

This may be a realistic assessment of consumers' behaviour and policy makers' attitudes, but it implies an evolution even weaker than 'Business As Usual'.

The Energy Directorate (DG XVII) of the European Commission stated in 1992:[6]

> Energy intensity will improve at a rate of some 1% per annum (to 2005). Improving further on this trend will require additional efforts by both the private and public sectors.

So here are two intergovernmental organizations expressing a view on the likely prospects that have, as members, many governments which have signed and ratified the FCCC. Clearly, if their judgements are correct on energy intensity reductions it is most unlikely that anthropogenic CO_2 emissions will be held at 1990 levels by the year 2000 or long thereafter. This is no doubt recognized by many of the staff in these institutions.

The WEC Commission, by contrast, was not prepared to consider a 'Business As Usual' evolution, particularly as this is incompatible with a range of stated and widely agreed goals. As the WEC Commission put it:[7]

---

5 IEA, 1994, p. 34.

6 EC (DG XVII), 1992, p. 19.

7 WEC, 1993, p. 245.

> Action needs to start now, on a rapidly widening multi-national basis. Much greater emphasis and real achievements are needed on energy efficiency.

The WEC regards the years 1995-2020 as the key transition period, hence the need for action *now*. Money should be spent to the greatest effect, rather than spending large amounts to achieve small incremental gains in those countries with the highest environmental standards and practices. This is *not*, however, to be used as an excuse for inaction on remaining pollution or where large efficiency gains can be made.

But it should be remembered that climate change is widely seen as of lower priority than local pollution and squalor. Furthermore, it has to be recognized that uncertainties abound over climate change. One just needs to consider the uncertainties that currently surround the carbon cycle (table 2.9). Inevitably, there will be pressures to delay action on climate change mitigation for as long as these uncertainties persist, despite the clear wording in Article 3 of the FCCC.

Table 2.9. Uncertainties in the Carbon Cycle Balance.

	1980s Mean Values (GtC/Year)	1850-1989 Total (GtC)
CO_2 Concentrations	3.4 ± 0.2	141 ± 10
Ocean Flux (F)	2.0 ± 1.0	111 ± 56
Industrial Resources (I)	5.5 ± 0.5	213 ± 20
Net Land Use Changes	1.6 ± 1.0	123 ± 40
Biomass Changes (+F-I)	-0.1 ± 1.1	39 ± 60
Apparent Budget Imbalance	-1.7 ± 1.5	-80 ± 72

Source: IPCC, 1994; Wigley, 1993.

Continuing uncertainties surrounding climate change also mean that the benefits of action specifically directed at climate change mitigation or adaptation are currently unknown, while uncertain costs will be incurred beforehand which could probably be justified on other, robust grounds. With action directed across a broader front than climate change mitigation, the costs specifically attributable to climate change mitigation could well be modest.

This position is perhaps a little more cautious than, but not very different from, that expressed by the World Bank in its 1992 World Development Report:[8]

8 World Bank, 1992.

> Bringing together the various estimates of economic costs and benefits leads to a simple conclusion: the balance of the evidence does not support a case of doing nothing, but neither does it support stringent measures to reduce emissions now - the costs are too high in relation to the prospective benefits.

The reason for the greater caution is that it is probably impossible to ascertain whether the (uncertain) costs are 'too high' in relation to (currently unknown) prospective benefits. The questions raised are fairly simple: what are the costs if local and regional pollution are also alleviated?; if basic needs are increasingly satisfied?; if energy efficiency is raised?; if cleaner energy conversion processes are introduced?; if more renewable energy is provided?; and if finance and technology flow in situations where they might not otherwise have been available? Furthermore, for a host of reasons given in this contribution, the WEC Commission did not accept the view of some prominent economists, especially in the USA, that action should be delayed because costs of mitigation will thereby become cheaper. This position may also reflect a greater enthusiasm among academics for discussing social discount rates and time-preferences than is evident among practitioners.

JI has been justified on narrower cost grounds, that the cost of x tons of GHG emissions saved or sequestered by a JI project in the host country is lower than can be achieved by action in the donor country. Indeed, it has been claimed that cost-effectiveness is the main rationale for JI.[9] However, if neither the benefits nor the costs of climate change mitigation are known, it must surely be questionable whether the difference between two uncertain sets of costs for currently unknown benefits is a prudent basis within the minimum regret criterion. Set within the wider agenda they could offer a great comfort factor.

In short, to avoid delay, to take action and carry out schemes which can be justified on many solid grounds, to meet the diverse priorities of countries around the world, and still to remain within the broad principles and objectives of the FCCC requires a broadening of the agenda for JI.

7 Rational Goals and Criteria for JI Schemes

Although a more detailed analysis on the criteria for the application of JI will be provided by the contributions in part III of this volume, this section will already make some recommendations on the setting up of JI Schemes.

---

9 e.g. in Kuik et al., 1994, p. xii.

The rational goals for JI Schemes would seem to be:

- acceleration of energy efficiency improvements;
- a programme for reducing subsidies, raising tariffs to cover costs, cutting transmission and distribution losses, and introducing of metering;
- encouragement of cost-effective renewable energy, provided adverse local environmental impacts are avoided or minimized;
- reduction of harmful emissions having local, regional or potential global effects;
- encouragement of local suitability and provision;
- support for international, joint co-operative ventures;
- a means of achieving early action;
- satisfaction of the need for precautionary measures of a 'minimum regret' kind; and
- avoidance of projects that could burden developing countries, and indeed other countries, with heavy external debts.

Suitable JI criteria would seem to be:

- ensuring the efficiency of expenditure in relation to emissions abatement;
- an equitable distribution of gains, to be assessed on a broader basis than just mitigation of GHG emissions or credits thereon;
- effectiveness in relation to policy goals, which need to be framed more broadly than climate change mitigation to encompass the priorities of developing countries and the economies in transition in relation to more local problems;
- adjustable to changing priorities and needs, which the broader agenda suggested here will support;
- simple and transparent to identify, arrange, administer, monitor and verify; and
- minimal barriers to speed and range of relevant project development.

There will be a need for a clearing-house to screen, monitor and verify projects. This should be done with minimal bureaucracy and privileged position-taking. One role for the clearing-house will be to encourage efforts directed at assisting the most needy, most vulnerable, and the worst instances of pollution and inefficiency. It is unlikely that these arrangements can remain entirely voluntary, given the requirements for verification and monitoring in recipient countries and for scrutiny of performance in donor countries seeking offsets. And, in keeping with the whole thrust of the WEC Commission's findings, when a broad appraisal of the full range of energy-related environmental problems plus potential efficiency gains is taken into account, it will be neither necessary nor desirable to phase-in JI over 10, 20 or more years or delay the introduction of JI Schemes.

The FCCC has, however, raised some matters of considerable concern in relation to JI criteria, screening, monitoring and verification. Thus the FCCC gives to the Conference of the Parties (CoP) the key role in establishing and reviewing JI criteria (Article 4.2(d)). Article 9 of the FCCC provides for a subsidiary body for scientific and technological reporting to the CoP which, though thankfully multi-disciplinary: 'shall comprise government representatives competent in the relevant field of expertise'. This subsidiary body is to identify innovative, efficient and state-of-the-art technologies and know-how, and to advise on promoting the development and transfer of these technologies. Article 10 provides for a subsidiary body for implementation, also comprised solely of government representatives.

This is all rather breath-taking for business people with direct experience in this field: experience of international investment, of technological innovation and diffusion, and of the hazards and problems of governments and government agencies trying to 'pick winners'. The analysis and conclusions of the WEC Commission pointed to the need for governments to set frameworks, and then let industry and commerce take the necessary decisions principally guided by market forces. The FCCC gives the impression of vulnerability to maximising bureaucracy, providing jobs for government representatives to the exclusion of others with practical experience, seeking to 'pick winners', and lacking a basic understanding of the operation of markets and international investment flows. The CoP would seem to be an unwieldy and inappropriate vehicle either for the purposes of a clearing-house, or for overseeing one. The need to attract wider practical experience and know-how into a small, dedicated institution is obvious.

8 Conclusions

The main conclusions from the foregoing can be brief:

- JI offers a means of early action;
- JI provides finance and technology;
- JI criteria can and should satisfy a range of efficiency, energy availability, and environmental goals;
- JI can place money where it has the biggest benefits;
- JI means international co-operative action; and

- JI should not get blocked by narrow cheating, chiselling and measurement concerns.

It may be objected that this is neither the definition nor the purpose of JI as given in Article 4.2(a) of the FCCC. However, this Article also makes reference to the need to take into account different starting points, economic structures, resources bases, and the maintenance of sustainable economic growth. The broader approach to JI advocated here, reflecting the WEC Commission's findings, takes into account those wider considerations and should not cause concerns about loss of sovereignty to be a real issue.

But it is of critical importance that in widening the criteria for cooperation the temptation to hand even broader powers over to the Conference of the Parties to the FCCC is resisted. As has been observed: 'Such efforts for a wider interpretation do not deserve support'[10]

However, conducted in the appropriate way and addressing generally agreed priorities, JI could prove for many countries the single most important international vehicle for action on sustainable development and environmental mitigation in the key transition period, 1995-2020.

---

[10] Kuik et al., 1994, p. 6.

References

European Commission (DG XVII), 1992. *Energy in Europe: A View to the Future*, September, Brussels/Luxembourg (ISBN 92 826 3665 8).

International Energy Agency, 1994. *World Energy Outlook*, OECD/IEA, Paris (ISBN 92 64 14074 3).

IPCC, 1994. *Radiative Forcing of Climate Change; CO_2 and the carbon cycle*, IPCC WGI Report.

Kuik, O., Peters, P., Schrijver, N., 1994. *Joint Implementation to Curb Climate Change*, Kluwer, Dordrecht/London (ISBN 0 2923 2825 6).

Wigley, T.M.L., 1993. *Balancing tne Carbon Budget; implications for projections of future carbon dioxide changes*, Tellus 45B, pp. 409-25.

World Bank, 1992. *World Development Report 1992: Development and the Environment*, IBRD/Oxford University Press, Washington/New York (ISBN 0 19 520876 5).

World Energy Council, 1993. *Energy for Tomorrow's World*, Kogan Page, London (ISBN 0 7494 1117 1), St Martin's Press, New York (ISBN 0 312 10659 9). Also available in French, Spanish, Portuguese, Arabic and some other languages.

World Energy Council, 1994. *New Renewable Energy Resources: A Guide to the Future*, Kogan Page, London (ISBN 0 7494 1263 1).

3. Carbon Dioxide Emissions

A cost-effective approach

CATRINUS J. JEPMA and CHE WAH LEE*

1 Introduction

Abating greenhouse gas (GHG) emissions on a national level involves substantial investment efforts, though part of these may be regained soon.[1] On a global level, the costs of the available options are likely to differ significantly between countries/regions. To proceed with greenhouse abatement without paying attention to these (large) cost differences would be to forego some of the most attractive opportunities: large cost savings can be made if pursued in a concerted action.

This contribution will go into some issues of an integrating assessment study of response options. The study is a simulation approach based on an integrating response framework. A few observations derived from the analysis will be discussed in order to sketch the scope for JI.

The study will not go into (normative) issues of how to implement a concept of JI or what criteria to observe for the distribution of costs.

2 Simulation and Sensitivity Analysis

In recent years, many studies were conducted to get a better insight into the costs of reducing GHG emissions. Amongst these are those that aim at constructing supply curves for global warming abatement measures.[2] Usually they are detailed technology based models, denoted as *least-cost greenhouse planning*, *integrating approach*, *bottom-up* or *engineering-economic* models. These models offer a methodology for comparing different technical options for GHG abatement. The attempt is to identify

---

* Catrinus Jepma is professor of economics and associated with the University of Groningen, The Netherlands, and Open University, The Netherlands. He is convening lead author of writing team 7 of the IPCC WG III Second Assessment Report. Che Wah Lee is graduate of the University of Groningen.

1 Whether or not net negative cost abatement options exist is subject to controversy in the so-called top-down versus bottom-up modelling debate. This however, is no subject of the present paper. For an extended overview, see e.g. Grubb et al., 1993, or UNEP, 1992.

2 To date, little can be said about the cost-effectiveness of adaptation options. Therefore, the analysis is usually restricted to mitigation options only. So it will be in this paper.

C. J. Jepma (ed.), The Feasibility of Joint Implementation, 57–68.

the optimal set of cost-effective options given a specified emission reduction target. This results in a priority scheme, ranking the options available according to their cost-effectiveness.

This chapter in essence follows the same line of reasoning. However, it does not pursue the same degree of detail in supply side and end-use options. The focus is on mitigation options as an aggregate of measures of the same type, rather than on the individual mitigation technologies. In the analysis four features stand out:

- comprehensiveness;
- a global perspective with regional distinction;
- use of cost functions instead of point estimates; and
- simulation and sensitivity analysis.

Comprehensiveness

Since CO_2 accounts for more than 50% of the total greenhouse warming potential of all GHGs and since little is known about the cost-effectiveness of measures concerning other GHGs, most attention is generally paid to CO_2 only. This is true for the present study as well.

However, any truly comprehensive analysis of global warming should allow for other GHGs as soon as additional data are available. Here, a few options concerning two other GHGs are considered, though still very tentatively, viz.: a switch-over of HCFCs to other CFC-substitutes; and the effect of methane leakage.

A global perspective with regional distinction

Whereas many studies have been conducted on the national level,[3] the focus here is on the global level, though not for the purpose of a single global outcome, but for an interregional comparison. Here, for the moment, only a few regions are distinguished.

Use of cost functions instead of point estimates

Until now, virtually all studies use *point estimates* representing the *average* analyzed/levelized costs of the distinguished options. That is, costs of a single measure are assumed to be roughly constant along the potential estimated. Subsequently, the options are ranked in order of cost-effectiveness, thus providing a stepwise increasing cost function (or blocked diagram) across the set of measures. Figure 3.1 illustrates this idea.

[3] For a detailed, well-based bottom-up study on the US see e.g. Rubin et al., 1992, on the UK see Jackson, 1991, and on several countries see Mills et al., 1991.

However, in reality costs are likely to rise not only for the aggregate of all mitigation options, but for each single option as well. In other words, costs per option are *not* likely to be constant, neither in terms of scale of application, nor in time. Therefore, using a single point estimate for the estimated potential of a measure may seriously fail to describe true cost behaviour.

Figure 3.1. An illustrative stepwise cost function/block diagram.

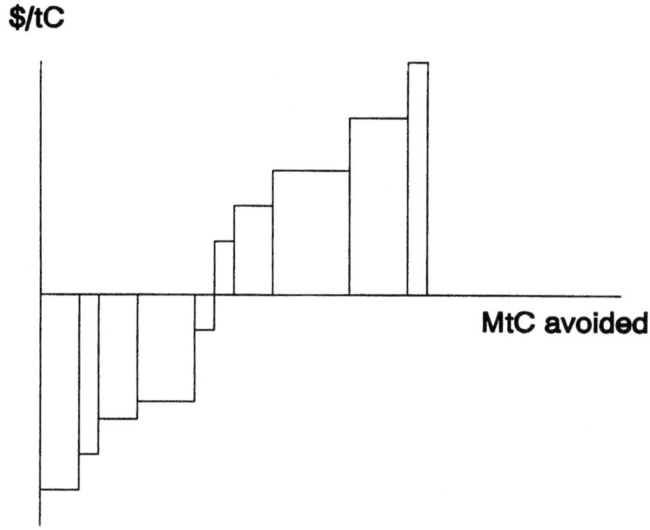

In forestry, cost estimating methodology is making progress towards explicitly accounting for increasing costs.[4] This development shows that point estimates may not be far from reality in the case of small areas, but are indeed unsatisfactory when considering larger projects and/or farther time horizons.

It is still to be seen whether cost estimating studies for other mitigation options will follow suit. However, there is no reason why the characteristic of increasing cost would not hold for mitigation options other than forestry. Therefore, in this chapter increasing cost functions are taken as a basic concept.

Simulation and sensitivity analysis

A truly sophisticated global assessment of mitigation options requires a host of data to a degree of detail that cannot yet be disposed of. For the present analysis, rough-and-ready estimates had to be relied on. Therefore, pending additional data, the

[4] *Ibid*, Moulton and Richards, 1990; Adams et al., 1992; and Parks and Hardie, 1992.

analysis outlined here cannot extend beyond the reach of a simulation study. Since the uncertainty ranges of cost estimates are large, the major parameters have been varied, providing a kind of sensitivity analysis.

3 Inputs and Results

In finding the least-cost allocation of options across regions, cost functions (specified per option and region) are critical. However, with the notable exception of forestry, virtually no data describing such functions are yet available. For the purpose of this chapter, the shape of these functions was roughly gauged, using a few selected studies as a point of reference.

Two sets of cost functions were taken as input for the simulation, yielding two simulation versions. One is from an earlier study using Nordhaus[5] and McKinsey[6] and distinguishing five mitigation options. The other is an update using studies by Jackson, Mills and Rubin and extending the analysis with three options.[7] Both simulation versions distinguish three regions. These features are indicated in table 3.1. Tables 3.2a and 3.2b summarize the set of estimates describing the cost functions. They clearly show costs to differ between regions and to rapidly increase after the most attractive opportunities are exhausted. What is most striking is that energy conservation and efficiency improvement notably exhibit large negative net cost opportunities in the second set of estimates.

For both simulation versions, a so-called base case simulation was derived. Subsequently, parameters were *ceteris paribus* changed so as to yield a sensitivity analysis. The conditions for the base case are not necessarily best or most probable, but rather constitute a feasible starting point for parameter changes.

The base case simulations are shown in table 3.3. They show large discrepancies, resulting in a total cost difference of more than $300 bln. This is due to several causes, the most important of which are the substantial cost savings by energy conservation and efficiency improvement and by a fossil fuel switch.

---

[5] Nordhaus, 1991a.

[6] McKinsey, 1989.

[7] Jackson, 1991; Mills et al., 1991; and Rubin, 1992.

Table 3.1. Features of simulation versions.

Simulation version 1	Simulation version 2
Regions:	Regions:
1. OECD	1. OECD
2. Eastern-Europe	2. Eastern-Europe
3. ROW	3. ROW
Options:	Options:
1. energy conservation and efficiency improvement	1. energy conservation and efficiency improvement
2. fossil fuel switch	2. fossil fuel switch
3. forestry measures	3. carbon removal and disposal
4. renewable energy	4. nuclear energy
5. phasing out CFCs	5. renewable energy
	6. forestry measures
	7.* phasing out halocarbons (CFCs,HFCs,HCFCs)
	8.* methane
Major sources:	Major sources:
McKinsey (1989)	Jackson (1991)
Nordhaus (1991a)	Mills et al. (1991)
	Rubin et al. (1992)

* not in base case.

Some of the results of a sensitivity analysis are summarized in table 3.4. From the perspective of JI, one phenomenon is noteworthy: no other *ceteris paribus* change in the sensitivity analysis has as dramatic an effect as the event of one single region, i.c. the OECD, assuming all of the burden of a global emissions reduction target. This is true for both simulation runs. Total costs change from $266 bln. (50% CO_2 plus CFC target) to $118 (20%) or $749 bln. (35%) in the first run, and from -$42 bln. (40% CO_2 target) to $7 (20%), $91 (30%) or $211 bln. (40%)! From this observation, it seems reasonable to conclude that a unilateral reduction effort of this kind is rather inefficient from a global point of view. If somehow a multilateral effort can be accomplished in a JI agreement, cost savings can be enormous.

4 Conclusions

The simulation runs in this paper have shown that substantial uncertainties still exist on the most efficient set of measures across the globe. Many studies disagree on the sign and the level of cost-effective measures, leading to a wide divergence of total cost estimates. However, this does not imply that nothing of interest can be said any more in greenhouse debate and a wait-and-see stance is the remaining policy strategy to be advocated.

The analysis outlined here shows that a global reduction target is much more likely to be achieved efficiently when exploiting cost-effective opportunities around the world first, rather than by exhausting options within a single region. In accomplishing such a global effort, JI can therefore play a major role.

Table 3.2a. Estimates of marginal costs ($/Tc) - Version 1.

		Level of emission reduction (MtC)						
Option	Region	0-25	25-50	50-100	100-150	150-250	250-500	500-1000
energy saving & efficiency improvement	OECD	-8	10	25	30	45	150	1000
	E-Eur	-5	15	30	85	150	600	1500
	ROW	3	20	70	100	190	800	1800
fuel switch	OECD	10	20	35	60	90	180	1000
	E-Eur	5	10	15	25	40	100	400
	ROW	15	30	80	100	180	240	1200
forestry	OECD	20	60	110	250	500	1500	3000
	E-Eur	12	20	40	75	100	300	2000
	ROW	5	10	20	30	45	60	100
renewable energy	OECD	10	25	60	95	200	600	1500
	E-Eur	30	100	160	200	800	1400	3000
	ROW	35	90	140	220	700	1500	3000
CFCs	OECD	5	6	15	20	40	60	500
	E-Eur	20	45	80	180	800	1200	2000
	ROW	12	45	90	140	220	500	700

Table 3.2b. Estimates of marginal costs ($/Tc) - Version 2.

		Level of emission reduction (MtC)						
Option	Region	0-50	50-100	100-150	150-250	250-500	500-1000	1000-1500
energy saving & efficiency improvement	OECD	-250	-200	-170	-100	100	200	1000
	E-Eur	-275	-150	-30	10	200	400	1500
	ROW	-100	-30	100	200	600	1400	3000
fuel switch	OECD	-150	100	200	450	700	900	1500
	E-Eur	-180	120	240	500	800	1200	1700
	ROW	-110	150	200	600	700	1300	2500
removal & disposal	OECD	40	50	70	105	200	600	1300
	E-Eur	50	75	150	250	250	800	1400
	ROW	60	85	165	300	275	900	1700
nuclear energy	OECD	30	100	180	350	600	1000	1500
	E-Eur	40	120	190	400	700	1300	2000
	ROW	60	130	190	425	800	1500	2200
renewable energy	OECD	30	90	275	600	900	1400	3000
	E-Eur	25	65	140	240	400	700	1900
	ROW	15	25	75	140	200	500	1400
forestry	OECD	4	7	11	35	100	400	1500
	E-Eur	3	6	15	32	85	200	600
	ROW	5	10	20	30	42	50	70
halocarbons	OECD	2	4	7	17	40	60	500
	E-Eur	30	65	200	500	700	1300	3000
	ROW	12	45	90	220	500	900	1200
methane	OECD	-4	18	24	155	500	800	1500
	E-Eur	-50	-20	25	60	200	400	750
	ROW	-80	-20	20	40	75	140	500

Table 3.3. Base case simulations.

Version 1	*Level of emission reduction (MtC)*			
Option	OECD	E-Eur.	ROW	Total
1. energy cons. & eff. improvement	350	250	150	750
2. fuel switch	250	500	150	900
3. forestry	100	250	1000	1350
4. renewable energy	150	50	100	300
5. CFCs	500	100	150	750
Total	1350	1150	1550	4050
	Costs ($ bln.)			
	OECD	E-Eur.	ROW	Total
1. energy cons. & eff. improvement	22.3	21	9.075	52.375
2. fuel switch	14.5	31.375	10.125	56
3. forestry	7.5	16.55	72.375	96.425
4. renewable energy	8.625	3.25	10.125	22
5. CFCs	21.025	5.625	12.925	39.575
Total	73.950	77.8	114.625	266.375

Marginal costs in optimum: $150/tC.

Version 2	*Level of emission reduction (MtC)*			
Option	OECD	E-Eur.	ROW	Total
1. energy cons. & eff. improvement	250	250	100	600
2. fuel switch	50	50	50	150
3. removal & disposal	100	50	0	150
4. nuclear energy	50	50	0	100
5. renewable energy	50	50	100	200
6. forestry	250	250	700	1200
Total	750	700	950	2400
	Costs ($ bln.)			
	OECD	E-Eur.	ROW	Total
1. energy cons. & eff. improvement	-41	-21.75	-6.5	-69.25
2. fuel switch	-7.5	-9	-5.5	-22
3. removal & disposal	4.5	2.5	0	7
4. nuclear energy	1.5	2	0	3.5
5. renewable energy	1.5	1.25	2	4.75
6. forestry	4.6	4.4	25.25	34.25
Total	-36.4	-20.6	15.25	-41.75

Marginal costs in optimum: $50/tC.

Table 3.4. Sensitivity analysis.

Version 1	*level of emission reduction (MtC)*				*costs ($ bln.)*			
ceteris paribus change	OECD	E-Eur.	ROW	total	OECD	E-Eur.	ROW	total
Base case	1350	1150	1550	4050	73.95	77.8	114.625	266.375
em. target 35% CO_2+CFCs	1200	675	960	2835	53.45	25.3	51.625	130.375
em. target 45% CO_2+CFCs	1200	995	1450	3645	53.45	57.3	100.625	211.365
em. target 55% CO_2+CFCs	1705	1200	1550	4455	133.35	85.8	114.625	333.775
em. target 65% CO_2+CFCs	1850	1300	2115	5265	161.45	104.8	236.225	502.475
renewables x 1/2	1350	1150	1550	1550	64.6375	70.175	108.0625	242.875
renewables x 1/5	1350	1150	1550	1550	56.05	63.8	101.725	221.575
renewables x 1/10	1450	1250	1350	1350	56.6875	69.675	76.6125	202.975
renewables x 1/20	1450	1300	1300	1300	50.75625	65.1125	63.30625	179.175
fuel switch	1450	1050	1550	1350	68.7	47.1125	104.5625	220.375
forestry x 2	1700	1150	1200	1200	139.45	91.35	114.5	345.3
forestry x 5	1775	1150	1125	1125	156.45	102.25	143.5	402.2
forestry x 10	1775	1100	1175	1175	158.95	96.25	168	423.2
only OECD, target 20%	1620	0	0	1620	118.05	0	0	118.05
only OECD, target 35%	2835	0	0	2835	748.95	0	0	748.95

Version 2	*level of emission reduction (MtC)*				*costs ($ bln.)*			
ceteris paribus change	OECD	E-Eur.	ROW	total	OECD	E-Eur.	ROW	total
Base case	750	700	950	2400	-36.4	-20.6	15.25	-41.75
core of three	1150	750	500	2400	77.5	36.75	20.25	134.5
8 options, targ 60% CO_2	1400	900	1300	3600	-22.5	-21.35	18.1	-25.4
em. target 50% CO_2	800	750	1450	3000	-32.9	-17.35	43.25	-7
em. target 60% CO_2	800	850	1950	3600	-32.9	-9.35	79.25	37
neg. cons&eff. x 1/2	750	700	950	2400	-15.9	-9.175	18.5	-6.575
neg.cons&eff.+fuel sw x1/2	750	700	950	2400	-12.15	-4.675	21.25	4.425
forestry x 2	800	800	800	2400	-29.3	-9.2	21.85	-16.65
forestry x 5	1100	700	600	2400	7.5	-12	12.75	8.25
forestry x 10	1150	650	600	2400	18	-13.5	16.5	21
only OECD, targ. 20% CO_2	1200	0	0	1200	6.6	0	0	6.6
only OECD, targ. 30% CO_2	1800	0	0	1800	91.1	0	0	91.1
only OECD, targ. 40% CO_2	2400	0	0	2400	211.1	0	0	211.1

Appendix: Some methodological problems

Least-cost greenhouse planning, though in essence very simple, is not free from some methodological problems concerning:

- time horizon;
- choice of a base year or reference scenario;
- time discounting/choice of a discount rate; and
- fuel cycle emissions/energy penalty.

Time horizon
Reducing GHG emissions on a global scale can obviously not be accomplished overnight. Therefore, any option considered in this analysis is unavoidably tied to a future time horizon. What time horizon to adopt is, however, far from obvious.[8] Moreover, in deriving cost estimates from different studies based on different time horizons it is difficult to accommodate for these disparities in order to avoid a degree of inconsistency to enter.

Choice of a base year or reference scenario
In calculating the costs of emission reductions, reductions are usually expressed in percentages relative to a base year or relative to a reference scenario. Likewise, costs are in any case to be considered costs to a base year or more appropriate to a reference scenario, reflecting future energy demand, energy intensity, carbon intensity and population growth. This implies that any forecast about future costs is based upon some explicit or implicit reference scenario. Since this simulation is partly to rely on these forecasts, a degree of inconsistency may arise as different forecasts are based upon different scenarios.

Time discounting/choice of a discount rate
There is no consensus whatsoever on the proper level of a rate with which to discount future costs. Different studies chose different discount rates. The influence of a discount rate can be far-reaching. In this study no explicit discounting has been done.

Fuel cycle emissions/energy penalty
Ideally, all fuel cycle emissions should be included, that is emissions in mining, processing and transporting the fuel as well as those in final combustion. For instance, transporting recovered CO_2 from the place of recovery to a place of disposal entails an energy penalty and thus a CO_2 penalty if done with fossil fuels. Here, no account is taken of these effects.

- For computing the set of least-cost options out of a host of non-linear cost functions, a non-linear computing programme is required. Unfortunately, given the large amount of variables, this was not available. Therefore, in this study the cost functions were formalized as discontinuous, linear graphs, in order to fit them in a linear programming model.
- The contribution of different GHG to global warming is expressed in CO_2 equivalent units, using the Global Warming Potential index (GWP) as a conversion factor. The GWP index is a measure expressing the amount (mass) of CO_2 leading to same degree of heat trapping as a one (mass) unit of a particular GHG, incorporating the (immediate) radiative forcing of the GHG and its atmospheric lifetime.

[8] Problems arise *e.g.* with respect to the Global Warming Potential measure. Adopting a short time horizon will give more weight to methane emissions relative to those of carbon dioxide. It will lead to policy decisions that favour short warming at the expense of the long term.

References

Adams, R.M., C.C. Chang, B.A. McCarl, and J.M. Callaway, 1992. 'The role of agriculture in climate change: a preliminary evaluation of emission-control strategies', in: J.M. Reilly and M. Anderson (eds.) *Economic Issues in Global Climate Change: Agriculture, Forestry and Natural Resources*, Westview Press, Boulder.

Blok, K., C. Hendriks, W.C. Turkenburg, 1989. 'The Role of Carbon Dioxide Removal in the Reduction of the Greenhouse Effect', in: *Energy Technologies for Reducing Emissions of Greenhouse Gases*, Proceedings of an Experts' Seminar - Volume 2, pp. 135-56, IEA/OECD, Paris.

Blok, K., E. Worrell, R. Culenaere, W. Turkenburg, 1993. 'The cost-effectiveness of CO_2 emission reduction achieved by energy conservation', in: *Energy Policy*, June, pp. 656-67.

Grubb, M. et al., 1993. 'The Costs of Limiting Fossil-Fuel CO_2 Emissions: A Survey and Analysis'. *Annual Review of Energy and the Environment*, 18, pp. 397-478.

Hendriks, C.A., K. Blok, W.C. Turkenburg, 1991. 'Technology and cost of recovering and storing carbon dioxide from an integrated gasifier, combined-cycle plant', in: *Energy - The International Journal*, vol. 16, pp. 1277-93.

Jackson, T., 1991. 'Least-cost Greenhouse Planning: Supply Curves for Global Warming Abatement', in: *Energy Policy*, Vol. 19, No. 1, Jan/Feb, pp. 35-46.

McKinsey Cy, 1989. *Protecting the Global Environment: Funding Mechanisms*, report prepared for the Ministerial Conference on Atmospheric Pollution and Climate Change, Noordwijk, November 1989.

Mot, E., et al., 1992. *Confining and abating CO2 from fossil fuel burning - a feasible option?*, Revised edition, TNO, Apeldoorn.

McKinsey & Company, 1989. *Protecting the Global Environment (Findings and Conclusions, Appendices)*, Ministerial conference on atmospheric pollution and climate change, Noordwijk.

Mills, E., D. Wilson, T.B. Johansson, 1991. 'Getting started, No-regrets strategies for reducing greenhouse gas emissions', in: *Energy Policy*, July/Aug, pp. 526-42.

Moulton, R.J., K.R. Richards, 1990. *Costs of sequestering carbon through tree planting and forest management in the United States*, USDA Forest Service General Technical Report WO-58, Washington, DC.

Nordhaus, W.D., 1991a. 'The cost of slowing climate change: a survey', in: *The Energy Journal*, vol. 12 (1), pp. 37-65.

Nordhaus, W.D., 1991b. 'To slow or not to slow: The economics of the greenhouse effect', in: *The Energy Journal*, vol. 101, July, pp. 920-37.

Parks, P.J. and I.W. Hardie, 1992. *Least-cost Forest Carbon Reserves: Cost-effective Subsidies to Convert Marginal Agricultural Land to Forest*, Washington, DC.

Rubin, E.S. et al., 1992. 'Realistic Mitigation Options for Global Warming', in: *Science*, vol. 257, July, pp. 148-9, 261-6, American Association for the Advancement of Science.

UNEP, 1992. *UNEP Greenhouse Gas Abatement Costing Studies - Phase One Report*, UNEP Collaborating Centre on Energy and Environment, Risø National Laboratory, Denmark.

4. Joint Implementation

A cost-benefit analysis

PIER VELLINGA and ROEBIJN HEINTZ*

1 Introduction

Compounded by the lack of practical experience with Joint Implementation (JI), evaluation of the actual implications for participants in JI projects, *i.e.* the advantages and disadvantages of JI, is difficult. The only certainty is that welfare effects of JI will definitely vary for various participants in climate policies, including investors in JI projects, partners, their respective governments, as well as non-participants (*i.e.* the global community).

This chapter will attempt to evaluate potential benefits and costs of JI on three different levels: the micro (project) level; the macro (national) level; and the global level. Differing (or even conflicting) interests and objectives of these participants will be established, and an attempt will be made to evaluate potential costs (risks) and benefits for participants, directly and indirectly involved in JI, and for non-participants.

2 Diverging Interests of Countries and Participants

Every country is unique. There are a number of structural differences, including economic and cultural differences which determine particular perspectives and objectives of countries. As has already been indicated by Pearce (see chapter 1), the Framework Convention on Climate Change (FCCC) recognizes different circumstances and capabilities of Parties, and therefore distinguishes three different groups of countries (Annex I, Annex II, and non-Annex I countries) each having their own commitments and responsibilities.

Both between and within these groups there are further differences. In developed countries, climate change strategies in general, and JI in particular, are probably determined by differences in costs of emission reductions, energy use and reserves, traditional relationships, *etc.*, whereas in developing countries these policies may be

* Pier Vellinga is Professor of environmental science and Director of the Institute for Environmental Studies of the Vrije Universiteit Amsterdam, The Netherlands. Roebijn Heintz was at the time of the conference an economist and researcher at the Institute for Environmental Studies, Free University Amsterdam.

C. J. Jepma (ed.), *The Feasibility of Joint Implementation*, 69–77.

determined by most urgent development needs, historical experiences, climatic and geographical circumstances (fragility of ecosystems), differing natural endowments *etc.* The focus in this chapter is primarily on agreements between OECD countries (Annex II) and developing countries (non-Annex I), although agreements between OECD and East European countries (Annex I) also merit attention.

JI is generally considered as an effective mechanism to mobilize new financial resources in the industrialized countries in order to explore new means for international cooperation. In addition to already existing or forthcoming official capital flows between governments, JI may also result in significant capital flows from private investors in OECD countries to their counterparts in developing countries. Furthermore, as was already mentioned by Jefferson (see Chapter 2), JI is considered by some as a renewed case for Official Development Assistance (ODA) focused on energy-related schemes, whereas for virtually all OECD-countries, little or no public support for increases in the ODA budget has recently been forthcoming. Finally, JI might also provide welcome additional financial means to the budget of the new multilateral fund for the environment, the Global Environment Facility (GEF), the size of which is limited compared to the demand and potential.

Not only private businesses, but also governments, and probably even non-governmental organizations (NGOs), can be investors in JI projects. In developing countries, companies and NGOs may also participate (as counterparts) in a JI agreement, but it is anticipated that governments will act as facilitators, and should in any case give the final approval to the project.

In the next sections, the interests of the direct participants will be evaluated at the *micro level*, where costs and benefits can be estimated at a project level. Even if the role a government plays in the initiation and implementation of a JI project is minor, governments of both investing and partner country should nonetheless be indirectly involved, as they have engaged in the commitments of the FCCC, and will have to report to the Conference of Parties (CoP) on all relevant climate change policies and measures.

At the *macro level*, there may be an effect of the sum of all individual JI projects on national economies, *i.e.* general equilibrium effect. Obviously, if a government has a direct stake in a JI project, the effects of the project at the micro and macro level will play an important role in taking relevant decisions.

At a *global level*, the accumulated effect of JI projects will affect all countries, *i.e.* the global community, including countries that do not participate in any JI project. For countries that have relatively little interest in participating in JI projects, probably as a result of structural differences mentioned above (different costs, historic considerations *etc.*), their position on JI will primarily be determined by the expected outcome of the cost-benefit analysis on the global level.

Obviously, expectations and objectives of governments, businesses and NGOs directly or indirectly involved in a JI agreement are not always congruent. Potential costs and benefits should be clarified at all three levels in order to get a clear understanding of different interests and incentives and to gain insight into the actual implications of JI on the various levels mentioned.

3 Micro (Project) Level

At a micro level, effects are examined from the perspective of the direct participants in the project. The main incentive for a private or government party to invest in a JI project is the obligation to reduce emissions, either through international (governmental) or through national (business community) burden-sharing systems, where the relevant party is asked to contribute its share in emission control.

Investment decisions will principally depend on a comparison between the project costs, including operational components, and the marginal costs of achieving these obligations in the home country. To be no worse off, the partner in the host country will claim at least compensation for the project costs. However, this partner will only have an incentive to participate in a JI project if the project brings additional (side-) benefits, such as improved knowledge and skills, and/or a share in the net economic benefit (financial or other compensation).

In the absence of directives from the FCCC, the ultimate benefit for either party will depend upon the outcome of negotiations between both parties, which in turn depends upon the value both parties attach to the benefits (including side-benefits) and costs that are taken into account in the negotiation process, and their respective bargaining positions. In a 'win-win' situation, both parties benefit, *i.e.* investors face lower costs than in the home country in achieving an environmental objective, and partners in the host country receive a compensation on top of the project costs (in the form of side-benefits and/or financial compensation). The potential costs and benefits of direct participants in a JI project are shown in greater detail in table 4.1.

Table 4.1. Potential benefits and costs on the micro (project) level.

Micro level	Potential benefits	Potential costs/ risks
Investor(s)	- costs savings (by avoiding more costly options)	- transaction costs (identification of projects *etc.*)
		- project costs (capital, technology, operational components)
		- risk of project-failure; loss of capital
Partner(s)	- costs savings (increasing profit as a result of more efficient technologies)	- contribution to project costs (labour *etc.*)
	- knowledge and skills	- risk of non-performance of investor
	- share in net economic benefit	

4 Macro (National) Level

At a macro level, the effects are primarily of national governments' concern. However, it will be virtually impossible for governments to make an exact estimate of general equilibrium effects. From the perspective of the government of the investing country, allowing for JI will probably create national support for climate change policies, in particular within the business community. Less emphasis on investments in 'no-regret' options in the home countries, such as energy saving policies, may be an adverse effect of the project. Table 4.2 lists a range of potential benefits for investing and partner (developing) country governments.

Developing country governments are likely to select projects that comply with national development priorities, and bring a number of additional benefits to the national economy, such as a decrease in dependency on imported fuels and improved access to modern technologies (through increased diffusion and a decreased dependence on patents). On the other hand, JI projects may conflict with partner or host country views regarding their permanent sovereignty over their own natural resources, an argument particularly relevant to large-scale and long-lasting afforestation projects. Furthermore, it has been argued that developed countries will use up the cheapest opportunities to reduce emissions, leaving developing countries with the more expensive ones once they have committed themselves to emission reductions. However, this argument probably does not apply in case of energy projects, as technological development will likely continuously generate new CO_2 saving and energy efficiency options. In order to get compensation, governments of partner countries may want to claim a share of the environmental benefit of the project, *i.e.* the 'environmental credit'.

Developing countries have as yet very low levels of carbon emissions per capita, and no quantitative emission limits under the FCCC. However, all governments are aware of their development needs and so it is recognized that developing countries' total emissions will increase. Therefore 'environmental credits' could conceivably be banked for the future, and the FCCC could provide facilities for such an arrangement.[1]

Table 4.2. Potential benefits and costs on the macro (national) level.

Macro level	Potential benefits	Potential costs
Investing country	- cost effective climate policy (national economic benefits) - opportunities for exporters of environmental friendly technologies (enlarging consumers' market)	- less investments in home country (loss of employment opportunity) - less investment in technology development (could undermine national energy saving policy) - less abatement of other environmental pollutants
Partner country	- additional resources for priority projects - improved access to modern technologies, including sustainable energy resources - decreasing demand for imported natural resources; improving balance of payment - improved infrastructure employment opportunities - other environmental benefits - share in net environmental benefit ('credits' for 'banking')	- sovereignty aspects - potentially reduced emphasis on development of national technology basis - potential loss of cheap reduction options (future obligations) - reallocation of labour and other input

[1] A decision of a country to save (bank) credits for future use is dependent on its social rate of discount, which in turn depends upon the perception of long term costs vs. short term benefits, both from an economic and social perspective, including the justification of passing on higher prices to future generations.

5 Global Level

At the global level, the effects of JI on the global community amount to the sum of the effects of all JI projects. Table 4.3 lists potential benefits and costs on a global level. An important potential benefit to the global community is the capacity of JI to help developing countries find a development path that entails low carbon emissions, rather than follow the same wasteful path industrialized countries have kept in their economic development. Furthermore, JI may be considered as a tool to raise emission targets in industrialized countries, as illustrated in the phased approach by Vellinga and Heintz (see chapter 10). Alongside benefits mentioned, JI may also involve a number of potential costs or risks, some of which may be avoided by a proper design of criteria for JI. Practical experience with JI may either confirm or reject existing apprehensions. In addition, a number of technical questions will probably be resolved through experience with JI.

Table 4.3. Potential benefits and costs on the global level.

GLOBAL LEVEL	Potential benefits	Potential costs/ risks
Global community	- more ambitious reduction targets and broader participation	- reduce incentive to develop technologies
	- help developing countries find 'low-carbon emission paths' in early stages of their development	- risk of 'emission leakage' (nominal value of 'credit' contradicts with real emission reduction; increase of global emissions)
	- reduce global costs of climate change policies	- risk of 'export of sacrifices' (no change in existing consumption and production patterns in the North)
	- reduce risk of climate change	
		- JI funds compete with GEF/ODA

6 Remaining Questions

There are a number of remaining questions, crosscutting all levels, including: the questions of *additionality*, involving the problem of assessing 'the situation in absence of JI', *i.e.* the baseline. These questions include the following:

- Should projects which are commercially viable be eligible for JI?
- Can a strict distinction between existing and future ODA, ODA/GEF and JI funds be guaranteed, and is this necessary?

- How should the 'baseline' question be solved, i.e comparing levels of emissions with and without JI. It has been argued that tackling this question on the macro and global level (assessing national/global baselines) is virtually impossible, at least in the short term. In this respect, developing baselines at the micro (project) level is more feasible and practicable.
- The question of the net impact of JI on *technological development*. At the global level, the risk of a reduced incentive to innovate technologies that correspond to economic and social structures of industrialized countries has been introduced into the debate. However, it is legitimate to respond that this reduced incentive may be compensated by effects on the micro/macro level, such as increased diffusion of existing environmentally friendly technologies and incentives to develop technologies that are adjusted to specific circumstances prevailing in developing countries.
- Finding the correct *balance between activities in industrialized countries and JI activities*, in order to avoid maintenance of existing wasteful production and consumption patterns. The approach of 'dual commitments' in which industrialized countries indicate separately which emission reductions will be achieved by domestic policies and which through investments in projects abroad as suggested by Vellinga and Heintz (see Chapter 10), can be considered as an answer to this question.
- The question of *equity*. How can the system guarantee that the net benefits will be equitably shared?
- Practical questions, including the questions of measuring, reporting and crediting over time.

7 Balancing levels

In designing the JI system, division of the emphasis over each of these three levels comes into question. The micro level gives particular emphasis to environmental (emission reductions) and business issues, whereas on the macro level effects concern economic and development issues. At the global level, cooperation takes place between governments and includes equity issues (burden sharing). In discussions on JI, on the one hand, it has often been suggested that JI should primarily be considered as a tool to reduce emissions, putting emphasis on environmental effectiveness, *i.e.* the project-level. Following this argument, JI should be tested against micro criteria in the short term. On the other hand, it has been suggested that in due course, JI could evolve as a tool to restore balance between national economies and global equity. However, it is questionable whether - and to what extent - it is a wise policy to apply environmental investments for other purposes than solving environmental problems.

Earlier in this chapter it was pointed out that those countries not participating in JI will also be affected by JI, either positively (share in JI benefits, for example, in the form of controlled global emissions), or possibly negatively (for example, risk of 'emission leakage' - see Table 4.3). It is likely that these countries will put forward the potential cost/risks at the global level, including equity issues and a reasonable balance between activities in industrialized countries and developing countries. Ultimately the micro, macro and global levels will have to be mutually balanced from the start.

Reference

Kuik, O., Peters, P., Schrijver, N., 1994. *Joint Implementation to Curb Climate Change*, Kluwer, Dordrecht/London (ISBN 0 2923 2825 6).

5. Joint Implementation

A critical approach

BILL HARE and ARJETTE STEVENS*

1 Introduction

This chapter intends to add some critical comments to the preceding discussion of JI in general and its pilot phase in particular. The comments and the recommendations put forward are based on the position Greenpeace International has taken in the JI debate.

Greenpeace has grave concerns that the use of JI measures for emission credits prior to the development and implementation of real and effective national emission reduction commitments will seriously undermine the political and environmental effectiveness of the FCCC. In addition, significant concerns remain unresolved as to the real and equitable economic effectiveness of JI.

2 Environmental Effectiveness of the Convention

As stated earlier the 'ultimate objective' of the FCCC is to achieve 'the stabilization of GHG concentrations in the atmosphere' at a protective level 'that would prevent dangerous anthropogenic interference with the climate system' which is to be achieved 'within a time frame sufficient to allow ecosystems to adapt naturally to climate change, to ensure that food production is not threatened and to enable economic development to proceed in a sustainable manner'.

The current emission stabilization commitments for developed countries are not sufficient to achieve this objective of the FCCC. The first priority for developed countries must thus be to implement the current emissions commitments and to negotiate further emission reduction obligations to be adopted at the first CoP. The use of JI to avoid even the current insufficient commitments undermines any possible realization of the FCCC's objective.

Greenpeace believes that the overall intent of the FCCC and of Article 4.2(a) (the provision that relates to JI) is that all Parties involved in JI must have similar emission commitments and clear emission baselines.

---

* Bill Hare and Arjette Stevens are associated with Greenpeace International, Amsterdam, The Netherlands.

C. J. Jepma (ed.), The Feasibility of Joint Implementation, 79–85.

JI conducted between Parties that have dissimilar quantitative obligations - that is, between developed nations which have quantitative emission commitments and developing countries which do not have any such emission commitments or economies in transition (EITs) with declining emissions - threatens the long term environmental effectiveness of the FCCC.

Under these circumstances there can be no guarantee that aggregate global emissions would be reduced, since there will be no way of verifying whether the JI actions have reduced emissions in these countries below what would otherwise have occurred. The process would be open to abuse, including double counting and the shifting of emissions to other countries.

Emission limits obligations or binding baselines on all parties involved in JI would prevent double counting, fake additionality, counter productive strategic behaviour (for example waiting with measures because a worse situation could attract more projects, bad environmental policy could be rewarded this way) and the shifting of emissions to other locations.

JI for credit between Annex II countries and other Annex I countries (*i.e.* economies in transition - the former USSR and eastern Europe) gives rise to a similar class of problems. The FCCC defines developed countries to include those of eastern Europe and the former USSR. The latter economies now often have declining emissions as a consequence of economic collapse as well as previous excessively inefficient energy use and production practices. This situation gives rise to the potential for JI to be abused by countries.

In this context Greenpeace believes that the intent of the FCCC is to reduce emissions below the level that would otherwise have occurred. Where the 'business-as-usual' emissions of one EIT party to JI are likely to be below the 1990 levels over the period to 2000 (and for some period thereafter) the potential exists for JI to result in no real reduction in their aggregate emissions. Parties could claim emission credits from actions in EITs, whilst they, or other Parties, may be undertaking actions, not reviewable by the FCCC, that have the effect of lifting emissions from their declining level towards the 1990 level.

In other words, returning jointly to 1990 emission levels by an Annex II (OECD) country with increasing emissions and an EIT country with otherwise decreasing emissions does not guarantee real emission reductions and is not consistent with the intention of the Treaty (see also the question on 'additionality' raised by Vellinga and Heintz in chapter 4).

3 Political Effectiveness

Greenpeace believes that attempts by developed countries to use JI measures to obtain emission credits could be seen as an attempt to evade the obligations on developed countries to act first. Such attempts could seriously undermine the confidence of developing countries in the FCCC and ultimately reduce the prospects of their agreeing to binding emission limitations. Achievement of the FCCC's ultimate objective of stabilizing GHG concentrations would thus eventually be threatened.

The FCCC requires developed countries to take the lead in addressing the problem and to do this by the adoption of national measures to limit their own emissions. The FCCC also obliges them to assist developing countries by paying the full incremental costs of mitigating measures and by the transfer of technology and information.

It is not consistent with the intent of the FCCC that developed countries should receive credit for what they are already obligated to do. Furthermore, with developing countries likely to have to adopt quantifiable emission obligations in the future, global equity may not be served if developed countries have already used up cheap emission reduction opportunities in developing countries.

In view of these considerations, Greenpeace believes that industrialized nations should focus on assisting developing countries by stimulating projects that reduce emissions. They should not receive credit for these projects because they are already a part of their obligations under the FCCC. Greenpeace also believes that a new international institution is needed to support the transfer of know-how and technology to developing countries, and others, in order that they may limit the growth in their emissions.

4 Economic Effectiveness

To a significant extent the overall economic effectiveness of global moves to reduce GHG emissions will depend on the rapid development and deployment of energy-efficient and renewable energy technologies in the world's largest economies. Advocates of JI as a cost-effective device for developed countries overlook the potential for this approach to undermine the rate of technological innovation and market penetration in developed economies.

By diverting effort from national emission reduction programmes and the policies designed to achieve these, JI could slow down or stop the adoption of much needed improvements in the energy efficiency of many industrialized countries. In particular it could delay the adoption of measures desperately needed in the transport sector to protect human health and the natural environment in addition to the climatic implications of this sector.

JI is often advocated from the point of view of a particular economic sector heavily dependent on the production of energy from fossil fuels. Often ignored is the perspective that those sectors would economically benefit from more rapid global adoption of energy-efficient and renewable energy technologies.

JI could thus increase the costs of achieving deeper emission reductions in the future through the inhibition of technological innovation and commercialization in developed countries.

5 Guiding Principles

In order to enhance the environmental effectiveness of the FCCC and to demonstrate the commitment of developed countries to a fair and equitable regime for limiting and reducing CO_2 and other GHG emissions, JI projects should be conducted that are additional to existing obligations. Moreover, JI actions should only involve Annex I parties having quantitative emission reduction commitments and agreed, reliable baselines and JI should not be used for emission credit at this stage until real, legally binding emission reduction commitments have been achieved by developed countries.

A secure multilateral monitoring and verification system must be put in place before JI actions can begin, to ensure that the overall effectiveness of the FCCC is not threatened. A bilateral system without independent verification and extensive reporting requirements is open to abuse. JI projects should be reviewable by the CoP, and its designated subsidiary body, to ensure that operational criteria are being met. Registering and verifying GHG inventories, emission baselines, and cost-effectiveness of the project compared to national options in the donor country and other information on projects is vital for the ongoing development of the global response to climate change. In addition the World Bank dominated Global Environment Facility (GEF) has a conflict of interest in relation to the question of project quality and is therefore not suited as controlling institution. The CoP should retain complete authority over the process.

Greenpeace believes that before any detailed criteria for the pilot phase are to be discussed, the CoP should agree on the following guiding principles:

Additionality

JI should be additional to the obligations of the Annex II countries to transfer capital and technology to developing countries and to pay the full agreed incremental costs of measures taken in developing countries to mitigate climate change and should be additional to projects that would have been carried out anyway.

Equity

JI projects should be socially acceptable in the receiving country and a social impact procedure should be part of the initial phase of the project. Such projects should stimulate the development of the local technological structure and ensure positive local social-economic (capacity-building) effects or compensation for negative effects to ensure the support of the recipient country/community.

Participation and transparency

The whole process of the development of a JI project (definition, implementation and verification) should be open to consultation (and approval) of the local community, NGOs and other interested parties to ensure that all local social, economic and environmental costs and benefits are included. Where necessary resources should be provided for local NGOs to participate meaningfully in the consultation phase.

Local environmental effectiveness

Local environmental effects must be positive, with projects subject to environmental impact assessment procedures (to help prevent other environmental disasters resulting from hydropower and nuclear power).

Comprehensiveness

A full accounting for the activities in the host country by the sponsoring party should be included to ensure that the intent of the FCCC is being met.

6 Pilot Phase: Key Issues

JI of emission reductions is likely to play a future role as one mechanism for meeting the objective of the FCCC, but it should only cover a small part of the further reduction commitments of Annex I countries.

Reaching a common understanding between all parties on the nature of JI, and generating confidence in the integrity of JI is crucial if any of its future roles is to be successful. Any pilot or experimental phase will play a critical role in whether such confidence is built and must be developed with great care.

If the pilot phase is to explore effectively the future shape and role of JI in meeting the objectives of the FCCC, it must reflect as closely as possible the most likely criteria and operating arrangements of eventual JI. The conditions and guiding principles such as those mentioned above must apply for the pilot phase as well.

Pilot phase projects should be conducted within a multilateral institutional framework and within a common analytical framework to allow coherent conclusions to be drawn from the pilot phase. A badly designed or *laissez faire* pilot phase would probably lead to great difficulty in drawing generalizations and overall conclusions. If the International Negotiating Committee (INC) has not been able to do so prior to it, CoP 1 and its subsidiary bodies should move quickly to develop and recommend such a framework for adoption by CoP 2.

7 Recommendations

- Clear criteria for any pilot phase must be agreed before that phase begins. The pilot phase could not begin before the CoP reach agreement on this phase.
- Multilateral structures for a formal, *ex ante*, project assessment process by or on behalf of the CoP must be established before the pilot phase begins along with an agreed multilateral process of project review once the projects have been completed.
- Experimental projects should only involve parties with clear emission baselines and GHG projections using IPCC methodologies.
- No emission credit sharing arrangements should occur during the pilot period or be considered until CO_2 emission reductions goals have been agreed by Annex I countries.
- Pilot phase projects should be designed to fully reveal transaction costs for testing cost-effectiveness and to cover a sufficiently long period of time to enable an evaluation of longer term cost-effectiveness.
- Secure operational and review structures, including extensive reporting requirements, must be agreed and established beforehand. The JI operation should be managed by a subsidiary body of the FCCC.
- Projects must be assessed on a transparent basis via lifecycle analysis of the environmental, social, developmental and economic impacts according to agreed criteria and methodologies. These must be agreed upon before the pilot phase begins and reviewed on a project by project basis.

- The period of any pilot phase must be agreed and must allow sufficient time to review the success and difficulties with projects and institutional arrangements. A minimum period of ten years is appropriate for this phase.
- Projects should be confined to those which result in clear CO_2 reductions.
- Forestry plantations for sequestering carbon should not be allowed as JI on the grounds of methodological uncertainty, risk associated with these projects and the unacceptability of permitting GHG emissions to increase via a sink approach.

6. Joint Implementation from a Southern Perspective

LAYASHI YAKER*

1 Introduction

The FCCC has been the centre of serious negotiations even within the industrialized countries. The non-industrialized developing countries of the South participated in the FCCC, expecting to reap the environmental benefits that would be translated into ecological conservation, providing for more sustainable livelihood, not only for the urban few, but also for the rural majority. It therefore gives me great pleasure to present these thoughts on what I consider to be the African point of view, which I believe is representative of the Southern perspective of the idea of JI within the context of the FCCC.

2 An Overview

The South should be seen as a major determining factor in global environmental issues not just because of its population and geographical size but also because of the levels of underdevelopment and poverty which have made it virtually impossible for the countries of this part of the world to keep pace with the advancement of eco-conservation in the North. Most importantly, the fact that the environments of the two groupings are more complementary than replicative, has given emphasis to the idea of joint implementation of programmes to address climate change. Consequently, the problem as well as the means to address the issue of climate change still need to be re-defined to reflect the many activities in the management of terrestrial ecosystems that also induce climate change. Within this context, I wish to propose a framework of action within our aspirations under Agenda 21 to address them towards the achievement of sustainable development early in the twenty-first century.

The implacable progress of industrialization in response to progressively higher and more complex consumption patterns has been increasingly associated with environmental pollution and risks to human life. These changes in consumption patterns have not only been limited to the developed world; they have been exported to and have become endemic, at least among the elites, in the South as well. While not being

---

* L. Yaker is United Nations Under-Secretary-General and Executive Secretary of the United Nations Economic Commission for Africa (UNECA), Addis Ababa, Ethiopia. At the Groningen conference, the outline of this chapter was presented by Haile L. Thebicke (UNECA).

C. J. Jepma (ed.), The Feasibility of Joint Implementation, 87–97.

able to develop its industrial capacity and capabilities to cope with the new trends, the South has helped increase the prosperity of the North. The South must therefore be an integral part of the equation for the reduction of the negative environmental impact arising from such activities, including the levels of atmospheric CO_2 and other GHGs. Increased production in the South to satisfy growing demand has often also meant that cheap, easily accessible and, unfortunately more polluting forms of energy have increasingly been resorted to.

Fortunately, both in the North and in the South, environmental awareness has been increasing. Indeed, the momentum towards enhanced environmental awareness took on high and encouraging dimensions since the Stockholm Conference on the Human Environment in 1972, and was greatly promoted by the activities of the World Commission on Environment and Development (WCED). The United Nations Conference on Environment and Development (UNCED), within the context of which the FCCC was negotiated and adopted in Rio de Janeiro, June 1992, took this awareness to even greater heights.

It is now generally recognized that there is need for more organized strategies for ensuring the sustenance of the development process through sound environmental management. More and more, this has also been the subject matter of international dialogue at different levels. The numerous international conferences and conventions on the environment as well as the present momentum enhanced by the UNCED and implementation of the decisions taken at the Conference, either within the context of Agenda 21 or of the various Conventions, are testimony of this. Consequently, the present discussions on the Southern perspective on JI of the commitments of the FCCC are timely.

2.1 The South and emerging environmental issues

More often than not, the North has identified what should constitute the priority global environmental concerns. It initiates policies, develops and promotes the implementation of activities for the management of global environmental commons. The global approach has often been promoted by the North and readily accepted by the South in preference to an approach which is based on the appreciation of local, national, subregional and regional conditions. Still battling to cope with the complex mutations in the local environment, the ordinary man in the South is often pushed, both by his government and the North, into prioritizing environmental problems along Northern lines and approaches. This is partly due to the fact that environmental discussions have remained too much the monopoly of *Southern* governments and the North. Popular participation has not been encouraged and thereby, ample opportunities for the involvement of the people in addressing environmental concerns have been missed.

The environmental problems of the South appear to merit attention only when they are linked to the interests of the North. The Tokyo Declaration on Financing Global Environment and Development adopted by the Eminent Persons' Meeting held in Tokyo, Japan, from 15 to 17 April 1992 was clear about the dangers the world faced:

> The human future is at risk. Wasteful patterns of production and consumption in industrialized countries, together with pervasive poverty and population growth in developing countries, are leading to the destruction of the Earth's ecological base and intolerable levels of human suffering and deprivation. ...It is time to re-evaluate the thinking which underlies our present society. A new environmental ethic needs to be established, new value systems accepted and supported by the citizens at grass-roots levels, and new approaches towards development adopted, based on the principles of: strengthening the links between people, environment and development; adjusting our behaviour to the natural order that lies behind the limited and vulnerable ecosystems of the Earth; and sharing environmental space equitably among all countries of the world, and acting to meet the needs of future as well as present generations.

This spirit of partnership in global environmental management has been underscored at UNCED. The responsibility for containing, reducing and eliminating environmental damage must be clearly defined and accepted. The concept that these responsibilities must be borne by the countries causing it, in relation to the damage caused, has never been fully accepted by the North. The North is the repository of the sophisticated technology, expertise and resources that makes possible the high standards of living in that part of the world and elsewhere and is responsible for the lion's share of the high level of pollution and pollution risks that exists.

The North should therefore be largely responsible for eliminating these adverse effects on the environment and human health, particularly in the developing countries, where the exploitation of natural resources is largely for basic raw materials on which Northern industries depend. This fact appears to be largely ignored by the North. Although it might adopt conventions, protocols and agreements that are aimed at controlling pollution, the North still allows industries to produce and export products, which have been banned for domestic use, to the developing countries being fully aware that these countries do not have the appropriate technology for mitigating their negative environmental and human health effects.

The achievement of sustainable development in the South is the ultimate goal and shared responsibility of the global community and requires serious global cooperation. Some of the crises in the South, it should be emphasized, arise from the present economic and political regimes. These crises did not only derive from Southern economic structures, but also from adverse international economic conditions which have not been conducive to the development of the rural sector which is, in itself, the backbone of Southern economies. It is largely through the often harmful impact of international factors on Southern economies that the environment has progressively deteriorated with the effect that the rural populations in most of the Third World countries have become increasingly poorer and are no longer able to feed themselves and have to import food. Drought, desertification, deforestation and soil degradation have been the major environmental catastrophes that have plagued Africa, for instance. The poor development and utilization of other natural resources have also added a new dimension to the problem. If these trends are allowed to persist, most of the Third World countries would head for a state of increasing environmental degradation, a decrease in rural production, low production in all sectors, debt, underdevelopment and dependence on external aid.

It has been the general consensus that the success of the efforts to achieve sustainable development will primarily depend on a strong political commitment and continued consultations and cooperation, within and among the countries between the different interests and groups. This is why the policy orientations must be clearly spelt out and the legal, administrative and institutional framework be developed to ensure that all necessary sectors can be orchestrated to achieve sound environmental management for sustainable development.

Although all global environmental issues are of concern to the South, not all of them command the same priority interest. While addressing all of them, emphasis must be given to those that are of pressing urgency to the region. Priorities must be determined by the prevailing environmental conditions in the Third World.

2.2 Increased environmental awareness in the North: implications for the South

Despite growing awareness of environmental hazards, the South has not yet prepared itself the way it should. The raising of domestic environmental standards on manufacturing processes in the North, without extending this policy to domestic companies investing abroad, is not enough to curtail environmental pollution, as this will inevitably encourage manufacturers to transfer the 'dirty' stages of the product cycle to developing countries where regulations are lax or non-existent. Health and environmental regulations on the sale of products in the EU can also mean that products failing to meet the new specifications are simply exported. The wave of dumping toxic wastes and dangerous products in Third World countries will see a sharp increase under such conditions. While conscious of the various sources and implications of environmental hazards, the South presently lacks the capacity, and means to sustain it, to handle many of the negative environmental impacts of the development process, hence the rapid rate of environmental degradation that it experiences.

Inextricably, the North and the South belong to one global system with varying ecosystems but with inter-linked conditions that influence climate changes. However, in terms of technological development, there is an unequal relationship where the North has the advantage of greater technological advancement and development.

Within the context of the African Common Position on Environment and Development, African countries recognize that environmental policies must be based on a precautionary principle that should govern the development and use of science and technology for implementing environmental measures. This should be perceived through anticipating, preventing and attacking the causes of environmental degradation, even if firm scientific proof is still lacking. Arguments based on 'reasonable doubt' should not be used as an excuse for postponing measures to prevent environmental degradation or to address emerging issues such as climate change and global warming.

2.3 The issue of climate change

The problem of climate change has arisen from the fact that there is a disequilibrium between the emissions of CO_2 and other greenhouse gases (GHG) that contribute to global warming on the one hand and the capacity of the natural ecosystems to absorb them on the other hand. The assimilation of CO_2 from the atmosphere by green plants has drastically been reduced by the extensive removal of vegetation through deforestation without equal re- or afforestation, particularly in Africa, Asia and Latin America. The gains of the South from global strategies aimed at atmospheric CO_2 reductions must be in the effective contribution by the international community to its afforestation programmes as well as combating desertification, and general ecosystem rehabilitation. There are many reasons for expecting and promoting joint ventures so as to remedy the situation characterized by the unbalanced development and economic relations between North and South.

More often than not, the global environmental problems and their solutions have been identified by the North and are often readily accepted by the South in preference to an approach which is based on the local, national, subregional and regional conditions. Although all environmental issues concern the South, not all of them command the same priority interest. While addressing all of them, emphasis must be placed on those ones that are of pressing urgency to the development for the region concerned. Priorities must be determined by the prevailing environmental conditions.

3 JI from a Southern Perspective

3.1 The Southern perspective

Climatic change and the problem of the ozone depletion are no doubt issues of serious global concern, which require collective and imaginative remedial measures. As I have said earlier, the responsibilities for this sorry situation are shared. The South, like the North, has also contributed to this untenable situation, mainly through massive deforestation. However, the South is at cross roads - whether to concentrate on the implementation of present prescriptions to restore the ecological balance or to continue with the present rather forced environmental realities. While recognizing that the South should be vigilant about restoring healthy environmental conditions, all along we have to recognize that in the spirit of shared responsibility the North should demonstrate higher political will in appreciating the environmental concerns of the South. The international community should have this as its central consensus.

With the problem of global warming, the countries of the South are faced with even more daunting dilemmas, particularly regarding the production and use of fossil fuels, seen by many as a source of significant pollution. For the developing countries that produce and export oil, drastic reductions, or even the containment of production and use of petroleum, would inevitably lead to precipitation of serious foreign exchange constraints and, thus, to incalculable loss of capacity to address their environmental problems.

We can have guidance from the imaginative implementation of, for example, Article 4 of the FCCC which deals with the commitment of the Parties, and its paragraph 2, which provides, among other things, that the developed country Parties and other Parties included in Annex I commit themselves specifically to adopt national policies and take corresponding measures on the mitigation of climate change, by limiting their anthropogenic emissions of GHG and protecting and enhancing its GHG sinks and reservoirs. These policies and measures will demonstrate that developed countries are taking the lead in modifying longer-term trends in anthropogenic emissions consistent with the objective of the Convention.

This will also demonstrate the recognition by the Parties that the return by the end of the present decade to earlier levels of anthropogenic emissions of CO_2 and other GHG not controlled by the Montreal Protocol would contribute to such modification. Policies and measures will take into account the differences in the Parties' starting points and approaches, economic structures and resource bases, the need to maintain strong and sustainable economic growth, available technologies and other individual circumstances, as well as the need for equitable and appropriate contributions by each of these Parties to the global effort regarding that objective.

Article 4 further provides that the CoP shall, at its first session, review, in the light of the best available scientific information, an assessment of climate change and its impacts, as well as relevant technical, social and economic information. Based on this review, the CoP shall take appropriate action, which may include the adoption of amendments to the commitments in subparagraphs (a) and (b) of Article 4.2. It shall also take decisions regarding criteria for JI as indicated in subparagraph (a).

It is important to note that what had started out to be a convention of rigour ended up as a framework convention because of the many unresolved issues raised by both the North and the South. This is a reflection of the divergences in the understanding of the causes of the greenhouse problems and the levels of sharing responsibilities with regards to the means for addressing it. The commitments in Article 4 must not only be seen in the light of GHG emissions and alternative patterns of consumption that will reduce them, but also in terms of the complementarity of varied environments and development patterns, affordable alternatives and the capacity to cope with implications of the introduction of high-cost alternatives.

The negotiations for the FCCC were of special importance to the South because of the linkages between climate change and mutations in the natural ecosystems, particularly in relation to desertification, biodiversity loss, energy and forestry issues. During the negotiations the industrialized countries focused on the stabilization and reduction of emissions of CO_2 and other GHG; commitments as regards the provision of financial resources and technology to developing countries were directly and indirectly linked to this.

While it is still necessary to agree on the thrust and content of joint action to mitigate climate change, it is important that the spirit of the Convention should include programmes to eradicate poverty which is a major cause of consumption patterns that encourage environmental degradation.

The ultimate objective of the FCCC is to stabilize GHG concentrations in the atmosphere at a level that would prevent dangerous anthropogenetic interferences with climate systems. To this effect, Parties to the Convention are expected, among other things, to take precautionary measures to anticipate, prevent or minimize the causes of climate change and mitigate its adverse effects. The thrust of the Convention is on the drastic reduction of GHG. The gains of the South from such global strategies have to be pursued and supported from the realization that the health of the environment in the developing world may perhaps be crucial to the health of the global environment. Within this context, there is ample scope for not only a better articulation of the strategies, but also the promotion of joint ventures geared to that effect.

The South must therefore collaborate with the North, take advantage of available experience and the North's technical scientific and managerial capability. This will call for rational human behaviour to conserve the earth in the interest of all.

3.2 Pre-requisites for the South to benefit fully from JI

For Africa, and indeed the South, contribution to the reduction of atmospheric CO_2 lies in activities to increase and conserve vegetation cover, including such economic activities that are related to agricultural production. There is need to initiate the implementation of all programmes for: the rehabilitation and management of land resources, particularly afforestation, reversing desertification, management of species and ecosystems; strengthening the meteorological observation infrastructure; and promoting research and development in the area of climate change. Activities to be undertaken would include:

- the design of policies and policy instruments for building capacity, including the required infrastructure, for research in activities to mitigate global warming; strengthening Southern planning and management systems in collaboration with the appropriate national and international organizations;
- improvement, in collaboration with international organizations, of the functioning and capacity of national institutions in all areas of global warming and climate change, including related environmental laws, rules and regulations; launching awareness raising campaigns to promote the application of the results of research relating to mitigating global warming and climate change; and
- the establishment of machineries to promote technical cooperation in development and transfer of technology with respect to management of activities to mitigate global warming and climate change.

This would require:

- the mobilization of resources from national governments as part of their regular planning process as well as through external assistance;
- that governments, in collaboration with the national and international scientific community and with the co-operation of relevant national and international organizations, promote and support research and development programmes that will enhance understanding of the problem and application of research results within their national/regional context; and
- that governments, in collaboration with the national and international scientific community and with the co-operation of relevant national and international organizations, promote and support research and development, particularly extension activities, in order to enhance the human resource as well as institutional capacity in this field.

3.3 Major areas of activity

From the arguments advanced above and for JI to be realistic and ensuring the full participation of the South, one of the major areas of focus, as earlier indicated, must be the alleviation of poverty. This will enable the South to adopt the high technological levels of production which ensure reduction of the emission of CO_2 and other GHG. A pertinent area in this regard would be the enhancement of agricultural productivity. To mitigate greenhouse effects, the increase of productivity of agricultural land by soil fertility improvement appears several times cheaper and has a quicker effect on the atmosphere than the plantation of trees, a solution largely considered in international meetings. This is a key sector in poverty alleviation through the promotion of sustainable agricultural practices. While contributing to the reduction of atmospheric CO_2 the promotion of sustainable agricultural practices will also contribute to food self-sufficiency and a reduction in the foreign currency expenditures arising from food imports.

Secondly, JI of projects to increase land vegetation cover should be given due attention. This will address the problems of massive deforestation and consequent desertification. There should also be support for research and systematic observation as provided for in Article 5 of the FCCC. Focus should also be on the establishment of data-bases and related infrastructure to facilitate and ensure the collection of data on the levels of GHG emissions and the absorption capacities of the different ecosystems. Whereas the need for measuring the levels of emission in the South is pertinent, the requirement for a statistical data-base on the absorption capacities of the various ecosystems in the South is paramount. Investment in joint implementation of programmes and projects aimed at capacity building in the areas mentioned above is also crucial.

For effective JI, Africa and the South must therefore ensure that the relevant institutional infrastructure will come into being. For instance, African countries will need to initiate, as a matter of urgency, national studies which should include national inventories of GHG, strategies for reducing their emissions and enhancing reservoirs as well as the socio-economic considerations of climate impacts. African scientists will need more sophisticated computer systems as well as aircraft and balloons to enable them to handle problems of general circulation as well as ocean circulation models. These will also be needed to strengthen research and training in other areas such as regional and subregional climate modelling, agriculture and forestry and alternative energy technologies.

There is also need for consistent and systematic maintenance of data-bases for the retrieval of rainfall and temperature data of the African region so as to enhance the prediction of future climate patterns. The meteorological stations in member states will need to be reinforced in monitoring weather and climate changes. Institutions such as the African Centre of Meteorological Applications for Development (ACMAD) will need to be strengthened in order to enable it to collect meteorological data, process and analyze these data with a view to providing systematic and periodical information to users for application in agriculture production, water resources management and other human activities which depend on weather and climate.

4 Conclusion

JI and concerted global action for a future with common environmental security must be the goal of all developmental action. Within the context of the different regional positions on environment and development for UNCED, there is not only need for the recognition of the challenges that the world community is currently facing, but also for tangible efforts and activities to engender change. The North must be seen as determined to deliver its share of this highly desirable global strategy for the restoration of the health of the world environment. The South too, should go beyond the present frameworks of addressing environmental problems through the progressive application of science and technology in the diffusion of which the North should play a more vigorous role. This will naturally require meaningful adjustments in the individual and collective policies of the Parties.

The challenges faced by the countries of the South, notably in Africa, are no doubt immense. The countries of the South not only need to faithfully implement the necessary changes in their environmental policies, but also to perceive the environment as one of the arguments in the development equation. There is also need to engender significant changes in their policy orientations to transform and technologically upgrade their economies. They also need to significantly sensitize the communities in their countries on the advantages of environmental conservation and upgrading. Deliberate efforts on the part of governments and the people are the surest ways in winning the battle against environmental degradation. These efforts should be supported and complemented by similar moves by the countries of the North, which should be seen as complementary.

PART II

INSTITUTIONAL OPTIONS AND APPLICATION OF JOINT IMPLEMENTATION: MEETING SUPPLY AND DEMAND

7. Joint Implementation

Application and criteria

JULIAN ROY SPRADLEY, JR.*

1 Provisions of the Framework Convention on Climate Change

The Conference of Parties to the FCCC is directed to 'take decisions regarding the criteria for JI' at its first session in March 1995 (FCCC Article 4.2(d)). These criteria, by law, must implement the structure provided by the FCCC.

It is essential, therefore, to begin any discussion of decisions on criteria for JI with an understanding of the relevant provisions of the FCCC. As discussed in Part I and below, the FCCC authorizes Annex I Parties to implement policies and measures to mitigate climate change jointly with other Parties. This authority includes: any measure; any GHG source, sink or reservoir; any economic sector; and all Parties to the FCCC. Moreover, it applies to any time period, including prior to the year 2000.[1]

The FCCC requires all Parties to implement policies and measures to mitigate climate change:

- All Parties are required to implement 'measures to mitigate climate change' and 'measures to facilitate adequate adaptation to climate change' (Article 4.1(b)).
- Each Annex I Party is committed to 'adopt national policies and take correspon-ding measures on the mitigation of climate change, by limiting its anthropogenic emissions of greenhouse gases and protecting and enhancing its greenhouse gas sinks and reservoirs' (Article 4.2(a)).

The FCCC specifically authorizes Annex I Parties to 'implement such policies and measures jointly with other Parties' (Article 4.2(a)). These and other 'policies and

* J.R. Spradley is Counsellor on International Environmental Law with the firm of Spradley and Associates in Washington, D.C. Prior to establishing his firm, Mr. Spradley was Counsellor to the Under Secretary for Oceans and Atmosphere, U.S. Department of Commerce, where he served as the principal representative of the Department in the activities of the Intergovernmental Panel on Climate Change and the negotiation of the U.N. Framework Convention on Climate Change. Mr. Spradley also represented the Department at the United Nations Conference on Environment and Development.

1 See, *e.g.*, Statement of the US Delegation on Joint Implementation, INC Ninth Session, Working Group I, Geneva, Switzerland (February 7, 1994). The US intervention reiterated its view that the Framework Convention 'allows for projects with all countries, for all gases and sectors, in the period prior to 2000.'

C. J. Jepma (ed.), The Feasibility of Joint Implementation, 101–109.

measures to deal with climate change should be cost-effective so as to ensure global benefits at the lowest possible cost' (Article 3.3).

Each Annex I Party is required to communicate, by 21 September 1994 and periodically thereafter, detailed information on its policies and measures on the mitigation of climate change (Article 4.2(b)). These communications are to include policies and measures that are implemented domestically and policies and measures that are implemented jointly. In addition, each Annex I Party is required to communicate information on its projected anthropogenic emissions (Article 4.2(b)). Similarly, these projections are to include information on both policies and measures implemented domestically and policies and measures implemented jointly.

It is important to note that the commitments in Article 4.2(a) and Article 4.2(b) of the FCCC are, respectively, for each Annex I Party:

- to 'adopt policies and take corresponding measures;' and
- to 'communicate detailed information on its policies and measures... [and] its resulting projected anthropogenic emissions'.

These commitments do not create legally binding targets and timetables for reductions in emissions of GHGs. Therefore, no Party, including any Annex I Party or any developing country Party, is under any obligation to reduce its emissions of GHGs to any specific level by any specific year.

As discussed below, over the next century the substantial majority of global anthropogenic emissions of GHGs are predicted to result from activities outside the OECD. While each Annex I Party has a qualified 'aim' for the returning of its GHG emissions, individually or jointly, to their 1990 levels by the year 2000, FCCC Article 4.2(a) and (b), other Parties have no 'aim' for their national emissions. In the near term, the joint implementation of policies and measures to mitigate climate change is the most viable, if not only, mechanism in the FCCC for addressing emissions of these other Parties.

2 The Criteria for JI

There is no authority under the FCCC for the Conference of the Parties (CoP) to determine, as a part of its decisions regarding the 'criteria' for JI, whether or when Parties to the FCCC may engage in the joint implementation of policies and measures to mitigate climate change. As discussed above, these matters are settled by the FCCC. Any change to restrict Annex I Parties from the authority for JI provided under the FCCC would require an amendment to the FCCC and cannot be adopted as a 'decision' on the criteria for JI.

The mandate for the CoP with regard to decisions on 'criteria' for JI, as for all other policies and measures, is to determine the methodologies for accounting, reporting and verifying the limitation, avoidance or sequestration of GHG emissions from the joint implementation of policies and measures to mitigate climate change.

Accordingly, the CoP should take decisions on criteria for JI that incorporate the following:

- the criteria should include all 'other Parties;'
- the criteria should include all GHGs;
- the criteria should include all policies and measures;
- the criteria should facilitate 'cost-effective' policies and measures;
- the criteria should provide methodologies and procedures;
- all JI measures should be communicated;
- JI savings should be calculated; and
- information on JI measures should be separate.

The criteria should include all 'other Parties'

The criteria should include all 'other Parties', as authorized in Article 4.2(a). Based on the findings of the Intergovernmental Panel on Climate Change (IPCC), the implementation of measures to mitigate climate change in all countries, including developing countries, is essential to the FCCC. The IPCC estimates that, without policies and measures to mitigate climate change, anthropogenic emissions of GHGs will rise from their current level of roughly 6 Gtons of carbon to roughly 22 Gtons of carbon by 2100.[2]

Of the current global emissions, less than 50% now result from activities within the boundaries of the 24 states that are members of the OECD. According to the IPCC and other projections, if emissions in the OECD states were stabilized at current levels, global emissions in 2100 would amount to some 20 Gtons, and more than 85% of the emissions would result from activities outside the OECD (see Figure 7.1). Indeed, even if CO_2 emissions in OECD states were reduced to zero by 2100, global emissions still would increase to roughly four times the levels in 1990.

[2] Intergovernmental Panel on Climate Change, *First Assessment Report* (August 1990); Intergovernmental Panel on Climate Change, *1992 IPCC Supplement* (February 1992). See, e.g., IPCC Scenario IS92a.

The criteria should include all greenhouse gases

The criteria should include policies and measures on 'all' relevant sources, sinks and reservoirs of GHGs and 'all' economic sectors, as called for in Article 3.3. In order to be consistent with the FCCC, the criteria for JI should encourage the broadest possible range of activities.

The FCCC has as a stated principle that policies and measures to deal with climate change should 'cover all relevant sources, sinks and reservoirs of greenhouse gases...and comprise all economic sectors' (FCCC Article 3.3). Under the FCCC, Annex I Parties are authorized to engage in the joint implementation of policies and measures to mitigate climate change involving any GHG; any source, sink, or reservoir; any economic sector; and any other Party.

Figure 7.1. Global CO_2 emissions (billions of tons of carbon per year).

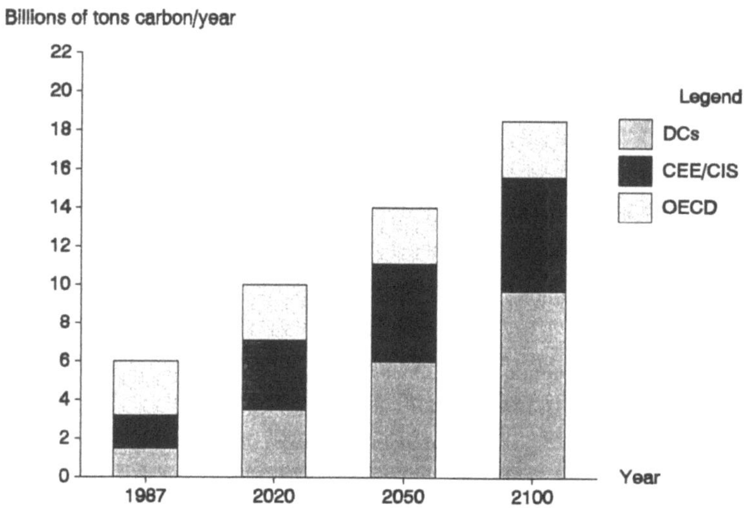

Note: The assumptions for the future projections are: that the developed countries would stabilize total CO_2 emissions at 1987 levels; that the formally centrally planned economies would have an increase in per capita CO_2 emissions of 50% by the year 2050; and that developing countries would have a doubling of per capita CO_2 emissions by the year 2050.

Source: Global Industrial and Social Progress Research Institute, 1991.

The criteria should include all policies and measures
The criteria should include all policies and measures that mitigate climate change independent of whether their primary objective is the limitation of GHGs, consistent with the *Guidelines for the Preparation of First Communications by Annex I Parties*, adopted by the INC at its 9th session.[3]

The criteria should facilitate 'cost-effective' policies and measures
The principles of the FCCC state that:

> policies and measures to deal with climate change should be cost-effective so as to ensure global benefits at the lowest possible cost (Article 3.3).

JI is one of the most cost-effective ways to achieve reductions in global emissions of GHGs. It is recognized that there is enormous potential for cost-effective GHG emission reduction through JI projects.[4] For example, studies of the former USSR identify reductions of 325 MMTC that are available by the year 2005, including 238 MMTC through cost-effective energy conservation measures with a negative net levelized cost.[5] A recent study in Poland identifies 16 measures to reduce emissions of GHGs that are 'feasible and profitable'. If implemented, these measures are estimated to reduce emissions of GHGs in Poland by more than 11 MMTC.[6]

These and other measures to mitigate climate change can also be implemented in developing countries. Given economic and other conditions, in the absence of JI, it is not likely that these measures would be implemented by the year 2000 in countries with economies in transition or in developing countries.

The criteria should provide methodologies and procedures
The methodologies and reporting requirements for the joint implementation of policies and measures to mitigate climate change should mirror the methodologies and reporting requirements for other policies and measures. On an interim basis,

---

[3] The Guidelines incorporate the IPCC *Draft Guidelines for National Greenhouse Gas Inventories*, INC, 1994.

[4] Federal Regulation, 1993.

[5] Battelle Pacific Northwest Laboratory, Global Studies Program, *Energy Conservation: The Main Factor for Reducing Greenhouse Gas Emissions in The Former Soviet Union* (January 1992); *Dynamic Cost Estimates of Carbon Dioxide Emissions Reduction in Eastern Europe and The Former Soviet Union: An Evaluation* (February 1992).

[6] The Polish Foundation for Energy Efficiency, *Evaluation of the Feasibility and Probability of Implementing New Conservation Technologies in Poland* (May 1993).

information on the joint implementation of policies and measures to mitigate climate change should be communicated consistent with other policies and measures using the Guidelines for the Preparation of Communications by Annex I Parties.

All JI measures should be communicated

The criteria should require that JI measures, and their effect on emissions, be included in national communications, as directed by Article 4.2(a) and (b). JI projects should be reported by each of the participating Parties. In its communication, each Annex I Party should separate its policies and measures to mitigate climate change implemented jointly from its domestic policies and measures. In its projection of emissions, each Party should separate emission reductions of JI measures from emission reductions from domestic measures.

The principal function of methodologies and reporting requirements is consistency in the communication of information. As for other policies and measures to mitigate climate change, policies and measures that are implemented jointly 'need not have as a primary objective the limitation of greenhouse gas emissions'.[7] Information on the 'motive' for JI measures is not required of an Annex I Party in its communication.

JI savings should be calculated

The criteria should provide methodologies for calculating savings from JI measures, in accordance with Article 4.2(c). The best technique for reporting JI is the 'project-by-project' approach. Respective whether the JI measure involves a source, sink or reservoir, a specific reference line can be developed for each project against which the savings attributed to the JI project can be calculated.

Information on JI measures should be separate

The criteria should provide procedures for communication of information on JI measures. The procedures should require that information on JI measures be communicated separately from communications on commitments in Article 4 for technical and financial assistance to developing country Parties.

Under the FCCC, JI is separate from commitments for technical and financial assistance to developing country Parties under Articles 4.3, 4.4 and 4.5. These provisions require, respectively, that Annex II Parties (OECD countries and EU):

---

7 INC, 1994, paragraph 16.

- Provide new and additional financial resources to meet the 'agreed full costs' incurred by developing country Parties for their national communications (Article 4.3).
- Provide financial resources, including for the transfer of technology, needed by developing country Parties to meet the 'agreed full incremental costs of implementing measures ... that are agreed....' (Article 4.3).
- Assist developing country Parties 'particularly vulnerable' to the adverse effects of climate change in meeting the cost of adaptation (Article 4.4).
- Take all practicable steps to 'promote, facilitate and finance' the transfer of, or access to, environmentally sound technologies and know-how to other Parties, particularly developing country Parties (Article 4.5).

3 A Pilot Phase

It would help the CoP in taking decisions on criteria for joint implementation of policies and measures to mitigate climate change if empirical information were available on these measures. This information could be collected during a 'pilot phase'. The criteria for a pilot phase should be based on the above-mentioned criteria for JI. In order to provide an empirical basis for subsequent decisions, the criteria for the pilot phase should be as broad and flexible as possible. The pilot phase should be limited to twelve months so that the CoP can take subsequent decisions on the criteria at its second session.

As discussed above, it is critical that the full potential for savings in emissions of GHGs through JI be realized as soon as possible. Therefore, JI measures that are now underway or are undertaken during the pilot phase should be included in national communications as measures to mitigate climate change.

It is important that empirical information on emissions savings from JI measures be considered by the IPCC in its Second Assessment Report in 1995 and by the CoP as a basis for the second review of the adequacy of commitments in 1998. To assist in this regard, the CoP should undertake an annual review of the available empirical information on JI. At its first session in March 1995, the CoP should begin this review with an examination of the potential for emissions savings from JI measures.

4 Conclusion

Over the next century, emissions of GHGs in developing countries and countries with economies in transition are projected to increase significantly. These emissions could be reduced markedly through the JI of cost-effective policies and measures. In the circumstances, the CoP should adopt decisions on 'criteria for JI' that facilitate the widest possible range of policies and measures to mitigate climate change.

The criteria for JI should include: all Parties; all greenhouses gases; all sources, sinks and reservoirs; and all economic sectors. The criteria should not arbitrarily preclude any policy or measure that can reduce, avoid, or sequester GHG emissions. Nor should the criteria restrict where or with whom these policies and measures can be implemented.

The criteria should provide an institutional structure for the joint implementation of policies and measures to mitigate climate change. In this regard, as a first-order priority, at its 11th session the INC should adopt methodologies and procedures for accounting, reporting, and verifying reductions in GHG emissions from JI. These methodologies and procedures should be based on the Guidelines for the Preparation of First Communications by Annex I Parties.

The CoP, at its first session, should adopt a twelve month pilot phase utilizing the above mentioned criteria. The pilot phase will provide empirical information for subsequent decisions by the CoP on the criteria for the joint implementation of policies and measures to mitigate climate change.

References

Battelle Pacific Northwest Laboratory, 1992. Global Studies Program, *Energy Conservation: The Main Factor for Reducing Greenhouse Gas Emissions in The Former Soviet Union*, January 1992.

Battelle Pacific Northwest Laboratory, Global Studies Program, 1992. *Dynamic Cost Estimates of Carbon Dioxide Emissions Reduction in Eastern Europe and The Former Soviet Union: An Evaluation*, February 1992.

Federal Regulation, 1993. 58 Fed. Reg. 66057, Washington, DC.

Global Industrial and Social Progress Research Institute, 1991. Report on global CO_2 emissions between 1987 and 2100.

IPCC, 1990. *First Assessment Report*, August 1990.

IPCC, 1992. *1992 IPCC Supplement*, February 1992.

Polish Foundation for Energy Efficiency, 1993. *Evaluation of the Feasibility and Probability of Implementing New Conservation Technologies in Poland*, May 1993.

INC, 1994. *Guidelines for the Preparation of First Communications by Annex I Parties*, INC IX, Working Group I, 7 February 1994, Geneva, Switzerland.

US Delegation on Joint Implementation, 1994. *Statement by the US Delegation on Joint Implementation*, INC IX, Working Group I, 7 February 1994, Geneva, Switzerland.

8. Joint Implementation

Institutional options and implications

PAMELA WREXLER, IRVING MINTZER, ALAN MILLER and DENNIS EOFF*

1 Introduction

From the analysis in Part I it can be concluded that JI describes a strategy through which countries may meet their commitments under the FCCC to reduce GHG emissions by financing emissions-reducing or sink-enhancing projects outside their own borders. The approach seeks to maximize the cost efficiency of emissions reductions on a global basis. Although the notion of JI was introduced during the Convention negotiations and the treaty references the term, details of how the regime will operate have not yet been decided.

In the Convention text no specific guidance on the meaning of JI was given, and no operational definition of the term could be agreed to during the negotiations. Since the presentation of the original JI idea during the climate treaty negotiations, discussion has focused primarily on criteria - what types of projects will qualify for credit and the mechanics of crediting reductions against specific obligations. A question attracting less attention is how the design of institution(s) established to manage the JI regime will determine the extent of emissions reductions achieved through JI projects and the degree to which JI supports the processes of sustainable development. Now, as the Intergovernmental Negotiating Committee (INC) prepares for implementation of the Convention and for the first meeting of the Conference of Parties (CoP), the importance of developing institutional arrangements for any future regime has become increasingly evident.

This chapter analyzes some of the institutional functions associated with the operation of a JI regime. Various conceptions of JI will be reviewed, and institutional functions that might promote desired goals will be proposed. Options for executing these functions - which range from a centralized bureaucracy to the participants themselves - will be identified and assessed, particularly in terms of the extent to which they promote the overall goals of JI and the FCCC.

---

* Pamela Wrexler, Irving Mintzer and Alan Miller are associated with the Center for Global Change, University of Maryland, Baltimore, USA; Dennis Eoff is associated with the US Environmental Protection Agency.

C. J. Jepma (ed.), The Feasibility of Joint Implementation, 111–131.

2 Objectives of Joint Implementation

The absence of a clear definition of JI in the Convention has led to the emergence of widely varying expectations and goals for the JI regime. JI programmes can be designed to serve many different purposes. Summarizing Part I, however, four key objectives for JI under the FCCC can be identified.

The first objective of any JI programme is to *identify and initiate cost-effective opportunities for reducing the rate of atmospheric build-up of greenhouse gases (GHGs).* Since the costs and extent of opportunities for achieving emissions reductions or sink enhancements vary among countries and across regions, a JI regime allows one country to underwrite relevant activities in another country at a cost per ton of avoided net emissions that is lower than that which could be achieved by the first country acting solely within its own borders. Proponents of JI hope that this regime will lead ultimately to inexpensive emissions reductions of a scale required to achieve the goals of the Convention.

The second objective of JI is to *support and encourage sustainable economic and human development.* By influencing the distribution of private capital, JI can help steer investment projects onto the path of sustainable development and emissions reductions. Depending on the percentage of emissions reductions abroad, JI also could be an important force in encouraging reduced emissions in developing countries. Ultimately, the success of JI - and indeed of the Convention itself - will be judged by whether it supports national development priorities consistent with sustainability on a global basis.

The third objective of any JI programme is to *encourage greater flows of private capital to applications that encourage local economic development through the dissemination of energy-efficient, emissions-reducing, and sink-enhancing technologies.* Indeed, part of the allure of JI is its potential to mobilize additional private resources, significantly more than if reduction projects were undertaken with governmental funds alone. This pool of private capital can improve access to environmentally-friendly emissions abatement technology, much of which is held by the private sector. A measure of the regime's success then, will be its ability to maximize the transfer of existing technologies, and to promote innovative partnerships to co-develop new technologies.

A fourth objective of JI projects is to indirectly *promote other policy objectives* that could be important *at the local, national and regional level.* For example, because emissions of local air pollutants often occur coincident with the emissions of GHGs, JI projects that reduce overall emissions through improvements in energy efficiency or the introduction of clean-burning, alternative fuel technologies will also reduce the emissions of the associated local air pollutants. Similarly, JI projects which enhance GHG sinks, either by expanding soil conservation programmes or afforestation efforts, will both slow the build-up of CO_2 in the atmosphere and generate local benefits in the form of increased food production or heightened forest yields.

3 Expectations for JI: Judging Performance of the Regime

Achieving an equitable and efficient regime for JI will be a complex undertaking, regardless of what institution or combination of mechanisms are designated by the Conference of the Parties (CoP). Realizing the potential of the regime will require a balancing of the interests of those who invest in JI projects and those who act as hosts for these undertakings. It will also involve trade-offs between the objectives of the JI regime described above. Indeed, the success of any proposed JI regime must be judged simultaneously along many dimensions. This section identifies five candidate dimensions along which such a regime might be judged.

Promoting ease of entry into the JI marketplace

JI can only make a significant contribution to the overall goal of stabilizing atmospheric concentrations if the market for JI projects is open and accessible to many players. For JI to be effective in achieving large net reductions in GHG emissions, many projects will be necessary, many players must be involved, and many technologies must be applied. No one type of project or technology has the potential alone to make a significant contribution to achieving the objectives of the Climate Convention. Any JI regime must be judged therefore, on the extent to which the proposed institutional structure promotes easy entry into the JI marketplace for new types of technologies and promotes innovative partnerships among enterprises in different countries.

Minimizing transaction costs

JI projects will be attractive only to the extent that they present a less costly means of achieving net emissions reductions than do the domestic opportunities available in the investing country. The JI regime will encourage investment in JI projects if the 'transaction costs' associated with mounting new projects are kept low. Elements of these transaction costs include search costs of finding hosts for investors and vice versa, information costs of designing and negotiating project terms, costs of accrediting the project and monitoring its performance, and costs associated with insuring for liability against project failure.

Facilitating the employment of environmentally-sound technology
Whether the technology employed in JI projects is new or mature, part of the allure of JI is its potential to promote the transfer and co-development of technologies that minimize the damage to both the local and global environment. These technologies - the backbone of a national or international strategy supporting sustainable development - are critical to reducing the risks of rapid climate change. The JI regime will be judged in part by its ability to promote such technologies as opposed to transferring conventional polluting technologies that have contributed to the current state of global risk.

Ensuring confidence among participants: fulfilment of financial obligations and achievement of project goals
Implementation and operation of JI projects are likely to continue over many years. In some cases, the full costs of construction and operation of the JI project will be provided by the investor in advance. In other cases, the investor may provide only capital and pay start-up costs and agree to make additional payments over the life of the project. Responsibility for ensuring that commitments are maintained could be left to project participants or could be managed by an international mechanism. In either case, the regime will not be considered successful unless financial commitments are met and projects are built and operated as conceived.

Another element of the JI regime's credibility among participants will be its treatment of failed projects. On the one hand, many JI projects will involve little or no experimentation with new technology. The risks of engineering failures in these projects will, therefore, be no greater than would be expected with any other international development project. On the other hand, part of the attraction of JI is that it may promote the development of new technologies and the co-development or refinement of existing ones. Some of the projects involving new technologies will inevitably fail to achieve their goals. One test of the JI regime will be how it deals with liability for such failures.

Ensuring credibility in the international arena: monitoring and verification
An important component of the JI regime will be provisions made for monitoring individual projects and verifying emissions. These evaluations will be necessary to ensure continued confidence of the international community in the credibility of the JI regime. Monitoring and verification could be the responsibility of participants in the projects. Alternatively, these activities could be conducted by local non-governmental organizations (NGOs), national governments, or an international mechanism implementing the JI regime.

Although several institutions exist that could perform monitoring and evaluation functions, the difficulties in executing these tasks must not be underestimated. The process of monitoring and evaluation of JI projects implies a complicated arrangement, especially if the JI regime expands to include many hundreds or even thousands of projects. The administrative complexities are complicated by a more fundamental challenge: the JI regime essentially will be called upon to measure invisible gases, and to assess the fulfilment of promises *not* to produce something. Whether the JI regime can credibly provide for comprehensive, systematic and long-term monitoring of project effects will be a critical element in judging the concept's success.

4 Institutional Functions within the JI Regime

The previous section laid out some of the dimensions upon which any proposed JI regime might be judged. This section examines the range of functions that might be necessary or desirable in a JI regime, and describes some of the possible institutional arrangements for executing these functions.

Since no operational definition of JI has been agreed to, the range of proposed functions that could be assumed institutionally is undetermined. Any or all potential functions could be assigned to some new centralized, international mechanisms designated to manage the JI regime, assumed by existing institutions including national governments, or performed by participants themselves. JI's management structure could thus fall anywhere on the institutional spectrum - from intervening hardly at all in JI project development and oversight, to directing the evolution of every JI project.

Accordingly, the discussion below does not presume a rigid sequence of institutional development; rather, it suggests points along the project development and implementation cycle at which institutional support may be necessary or desirable. The range of institutional options for executing suggested functions is also considered. Table 8.1 describes some of the functions that might be necessary to manage the JI regime.

4.1 Participants in the JI regime

There are two classes of participants in the JI regime:

- countries which are Parties to the Climate Convention; and

- enterprises that participate in JI projects.

The latter could include, for example, banks, private companies, NGOs and subnational entities. Enterprises will likely participate in projects that are clearly within their principal arenas of expertise, but the institutional role of national governments may vary considerably. A nation could create new domestic institutions or authorize existing ones to perform any or all of the JI functions listed above. Alternatively, governments could act only as rubber stamps for operations carried out by entities within their borders. Some nations, for instance, may choose to become involved with project development and implementation, while others could choose instead to concentrate on monitoring projects so that emission reductions can be claimed in the international regime.

Table 8.1. Potential Institutional Functions.

Function	Executor	Description
Identify the participants	Bulletin Board	provides investors and hosts a mechanism for finding each other
Facilitate the match	Broker	brings together investors and hosts; uses own judgment to propose good matches
Assist in project development	Advisor	provides technical, legal, engineering or economic advice
Certify the project	Approver	evaluates whether project meets goals of FCCC, CoP guidelines
Syndicate the financing	Packager	offers projects as investment instruments or bundles projects into portfolios, setting the price investors pay for participation
Guarantee project performance	Insurer	assumes responsibility for failed projects
Monitor project achievement	Watcher	monitors ongoing project performance and reports results to CoP
Verify project characteristics	Authenticator	verifies reductions and informs CoP
Resolve disputes	Adjudicator	resolves verification disputes; assigns liability (loss of credits) for failed projects

4.2 Facilitating deals

Identifying the participants

Because the assortment of potential players in the JI regime is not necessarily known to each other, a mechanism to link potential investors and project hosts was suggested early during the climate treaty negotiations. For example, a simple 'bulletin board' could act as a centralized collection point for information regarding potential JI projects and financing. Countries and entities could post and/or elaborate their project interests. Table 8.2 describes potential advantages and disadvantages of creating such a bulletin board.

Some type of linking mechanism benefits potential players in the JI regime by providing a point of access to information about JI project and financing opportunities. Given that the costs of searching for potential partners and obtaining information about potential projects might be burdensome to individual participants, reducing upfront search and information costs in this way could lower barriers to entry into the JI regime. Consequently, a greater number of entities - both recipients and investors - might participate, particularly those interested in small projects where search and information would constitute a larger percentage of total costs.

Table 8.2. Identifying Potential Partners.

If centralized among one or several institutions		if performed by individual players	
+	-	+	-
- reduces search and information costs for individual players ↓	- administrative costs of maintaining institution may be significant	- no centralized administrative costs	- potentially high search and information costs to individual players
- lower search and information cost reduces barrier to entry ↓	- discourages partnerships involving proprietary information		- differential distribution of information about project and financing opportunities
- may attract more players			

Regardless of the size or authority of an institutional 'bulletin board', its operation would incur measurable administrative costs. Under one scenario, these administrative costs could be diffused among all bulletin board participants by adding an average amount to the cost of each project. For participants without partners, the added cost to support an institutional mechanism which reduces search costs still is likely to be lower than if those search costs are borne individually. As the number of JI project opportunities increases, the per unit transaction cost of administration might be lowered further.

A mechanism for matching potential investors and hosts could exist in tandem with the development of partnerships formed without involvement with the centralized information mechanism. A strategy which required all JI opportunities to be publicly posted through some institutional mechanism could promote equitable distribution of information among potential participants. And adding entities which have had low search costs to the participants pool could lower aggregate search costs. But for participants who already had identified partners, a posting requirement might raise project costs and delay implementation. Posting might also discourage participation by entities which seek to keep project information proprietary.

Facilitating the match

Beyond the bulletin board function, some institution could be granted authority to facilitate the match between potential investors and project hosts. Conferred with this brokering function, the institution would use its own judgement to propose appropriate matches between investors and hosts. The institution additionally could be authorized to seek out host and investor participants. Table 8.3 summarizes potential consequences of institutionalizing an active brokerage function.

Table 8.3. Brokering Matches.

If centralized among one or several institutions		If performed by individual players	
+	-	+	-
- further reduces search and transaction costs to individual players, reducing entry barrier ↓ - identification of private sector entities channels expertise into regime ↓ - may increase credibility of regime	- may confer undue influence on a single institution - may inhibit development of innovative partnerships	- allows for unhindered associations among hosts and investors - entrepreneurs will bring innovation to the market, bearing the risks and the costs	- high identification and search costs

Like a simple bulletin board service, a mechanism for matching hosts and investors would carry administrative costs, but would reduce up-front costs to individual projects by allocating a portion of aggregate search costs among all participants. An active matching service might reduce costs further because it also would diffuse the bulk of information costs. Reduced costs resulting from institutionalized brokering would promote entry, and might also stimulate participation among a wider class of entities for which participation would not have been cost-effective if the expense of finding suitable partners was borne individually.

Creating or authorizing brokers to match project and investment opportunities might have advantages over a system where all deals were negotiated bilaterally. For instance, brokerage institutions could add an aura of credibility to the regime, promoting greater numbers and greater diversity of participants. If brokerages were able to seek participants, the character of the JI regime might be enhanced by the channelling of private sector expertise.

Conversely, institutions holding authority to match partners could gain undue influence in the JI regime. Brokers might demonstrate institutional biases in preferring particular types of technologies, with one result being that brokers ended up picking winners and losers. These biases could also limit the formation of creative forms of partnerships and the development of innovative technologies.

The brokerage function could be centralized, or assumed by several institutions. Permitting a monopoly in the matching and assignment of projects might result in more consistent arrangements, but the trade-off may be to increase delays in project implementation for more experienced players. Allowing for bilateral arrangements as well as for the creation or existence of multiple brokerage institutions with different structures would allow simple projects to be mounted quickly and allow for the formation of innovative partnership arrangements. The availability of options in matching project hosts and financiers is likely to promote participation by more and a wider variety of players.

4.3 Assisting project development

Among the components of JI project development are formulation of a concept, articulation of technical and legal elements, and analysis of costs. One option under the JI regime would be for individual participants to perform these tasks, that is, to develop projects themselves. An alternative scenario would be to centralize technical, legal and economic expertise in one or more institutions in order to assist potential players seeking to formulate JI projects. Table 8.4 describes some consequences of choosing one or the other of these options.

The value of institutionalizing some or all project development tasks lies primarily in the centralization of expertise. Without institutional assistance, participants themselves would need to become competent evaluators of technical, economic, legal and political aspects of project design. Any deficiencies in the ability of individual entities to adequately evaluate these elements of project formulation and design would increase the risk of project failure. But because institutions are likely to incorporate biases for particular technologies or designs, development of projects by individual entities could promote greater innovation.

Several existing institutions could execute some or all of the project development functions. Examples include private companies, national research and development laboratories, universities and private brokerage houses.

Table 8.4. Assisting Project Formulation and Development.

If centralized among one or several institutions		If performed by individual players	
+	-	+	-
- could provide high level of expertise to all players	- large bureaucratic institutions are often inefficient and slow to react to market conditions	- could promote a wider range of design concepts and technological innovation	- deficiencies in environmental expertise may increase risk of project failure

4.4 Packaging projects as investment instruments

Beyond facilitating matches and developing projects, management institutions within the JI regime could allow for the creation and oversight of new investment instruments. Under this approach, the JI institution would determine the cost of a project and set the price under which governments or private enterprises could buy a given level of emissions abatement. To take this approach a step further, the institution could package various projects in a portfolio, such that no individual investor would be tied to any particular project. Table 8.5 reviews the comparative features of bilateral deals and multilateral investment schemes.

A regime which packages JI investment opportunities could have several advantages over a market which only includes bilateral deals. Specifically, an institution which syndicates project investments could increase the attractiveness of JI to a wider set of participants. Such an institution would make the market significantly more accessible to small investors because virtually all up-front development costs would be borne centrally, and because small projects would no longer carry disproportionately large implementation costs. A syndicator could increase market accessibility for entities with interest in the JI market but without the capabilities of mounting emissions reduction or sink enhancement projects on their own. Syndication would provide potential project hosts with access to a broader capital base and thus access to more diverse projects than available under a straight bilateral system.

Because syndicated arrangements would diffuse technological and political risks among a pool of investors, they could attract participants from two important classes of potential investors: entities which would be too conservative to participate on their own in a new and rapidly evolving market, and entities which would not have the resources to support continued oversight of a JI project. This arrangement might cause some projects to lose most of their innovative aspects in order to secure greater participation among conservative investors. Another danger might be less rigorous project implementation than when hosts are responsible directly to individual investors.

One potential benefit of the investment syndication arrangement is the institution's ability to promote the goals of sustainable development. For example, the CoP might assist syndicators to steer investment toward environmentally sound and sustainable projects, potentially moving more towards these goals than if investments were judged only in terms of their potential for financial profitability. A centralized, institutionally-directed project development process might thus be more acceptable both to developing country governments and to NGOs which will likely have more access to project details under an internationally managed system than if projects are only negotiated on a bilateral basis. On the other hand, adopting selection criteria other than economic efficiency might thwart the benefits of the market and thus the full potential of JI.

Moving to a multilateral investment arrangement may hold some negative implications for efficiency, innovation and accountability. First, as with any institutional structure, the syndicator will carry administrative fees. Also, unlike bilateral arrangements where the host retains any difference between the negotiated price and its marginal cost of abatement, a syndicated scheme would leave that amount (a surplus) with the syndicating institution.[1] The price of emissions credits would not be set at a level equal to the least-cost emissions abatement in either case, but when the institution receives the surplus, there are further risks because the host's incentives to minimize abatement costs are diminished. Loss of the profit incentive, combined with the absence of any direct relationship between investors and hosts, could detract from the host's incentive to ensure the venture's long-term success.

---

[1] Because price is negotiated, the investment amount may exceed the true cost of the project. But regardless of who accrues the surplus, the project would be attractive to investors as long as the negotiated project cost was lower than the investor's cost of achieving reductions domestically.

Table 8.5. Characteristics of Bilateral And Multilateral Packaging Schemes.

Bilateral		Multilateral	
+	-	+	-
- encourages long-term relationships - maximizes entrepreneurial potential of individuals and private sector entities	- transaction costs borne by individual projects - may favour fewer countries with 'best' (cheapest or most extensive) reductions - parties must be competent evaluators of costs	- transaction costs diffused over entire portfolio - distribution of risk - hosts have access to greater capital pool - investors need not oversee project implementation - no long-term obligation to particular technology or entity	- overhead fee for syndicator - discourages long-term relationships - bureaucracy may not promote innovative projects or partnerships

Syndicated arrangements might also discourage long-term relationships among hosts and investors, with both positive and negative consequences. Without stable long-term partnerships, shared visions are less likely to develop. Thus, the willingness of investors to assume the risk of technological innovation might be much less than if they enjoyed stable long-term relationships with their partners. But with syndicated arrangements neither host nor investor would incur any long-term obligations to particular technologies or countries.

In the longer term, if private firms are encouraged only to buy reductions by investing in portfolios of projects, these firms may not take as strong an interest in the development of innovative and efficient projects. The JI institution, like other bureaucracies, may not be able to match the entrepreneurial and innovative spirit typically found with individuals and in the private sector.

4.5 Who could act as syndicator?

A variety of private and public sector institutions could assume a role in the creation of syndicated or multi-project investment instruments, including multilateral development banks, governments, NGOs, commodity exchanges, and private sector entities.

The Global Environment Facility

Recent proposals have suggested a role for the Global Environment Facility (GEF) in the JI regime.[2] While the concept has yet to be fully considered and operational details of the GEF itself have just recently been agreed upon, the proposal would complicate the role of the GEF if private funds and GEF grant funds were commingled. Some commentators have counselled, therefore, that JI and the financial mechanism should be held strictly separate.[3] Others have suggested that the GEF bureaucracy may not be able to respond rapidly to private sector proposals. On the other hand, this type of cofinancing could allow the GEF to leverage substantial private sector resources and amplify the impact of its own limited capital.

Acting as the financial mechanism of the Convention, the GEF has acquired substantial information and knowledge about some of the technologies and systems that could be used in JI projects. As a consequence of its experience in evaluating and underwriting such innovative projects, the GEF has become an important repository of current information relevant to the introduction and operation of JI projects. By emphasizing transparency in its operations, the GEF could become an active archive of this information, even if it does not assume a primary role in the JI regime.

The International Finance Corporation

The World Bank and its private sector investment affiliate, the International Finance Corporation (IFC), have suggested the establishment of a Greenhouse Gas Mitigation Venture Capital Fund.[4] The IFC proposal is still evolving but as currently envisioned, the Fund would mobilize both public and private capital to leverage primarily foreign private investment in projects which would cost-effectively reduce GHG emissions and are consistent with national sustainable development plans and programmes. The IFC is seeking financing for a feasibility study to determine the parameters of the Fund and IFC's role in it.

With its financing expertise and established relationships among players in significant sectors (especially electric power and forestry) the IFC might avoid the appearance problems raised by the commingling of GEF and private sector funds, while

---

2 See, for instance, Newcombe and de Lucia, 1993.

3 See, for instance, INC, 1993 ('Another important point to make is that investments in joint implementation should not be mixed up with the financial mechanism. JI should be held strictly separate...'); see also, Environmental Defense Fund, 1993.

4 Global Environment Coordination Division, 1993.

retaining the benefits of leveraging private sector resources while using public funds. The IFC also has expertise with assessing business risk, developing financial packages and forming consortia with foreign and local private investors in development projects. However, the IFC might discourage development of new technologies because its investors are likely to demand less risky, commercially demonstrated technologies.

NGOs

Although in principle non-governmental organizations could act as syndicators of JI projects, NGOs are less able to capitalize such enterprises. The credibility of NGOs as project *adelles* may be questioned by many investors. Nonetheless, in a few specialized areas - perhaps including the development of biodiversity reserves for commercial purposes - NGOs may be able to act successfully as syndicators. The expertise of these organizations in biological conservation activities and sustained yield management may offer the basis for developing credible JI projects that promote local development objectives and enhance GHG sinks while protecting biological diversity.

Any or all of these existing institutions or some newly created institution could assume a role in project syndication. And the existence of syndicated arrangements or multi-project investment funds need not preclude the formation of bilateral deals, either privately or through the use of a matchmaking institution. Keeping the JI regime sufficiently flexible to allow multiple types of arrangements is consistent with the opportunities JI provides for optimizing global costs and encouraging innovative partnerships between the public and private sectors. Regardless of how projects are facilitated, the consistency and credibility of the JI regime can be maintained by subjecting all projects to the same system of monitoring and evaluation.

4.6 Certifying project acceptability under the JI regime

The point at which a project is deemed acceptable to the international JI regime and the institution by which this judgement would be made could vary considerably. Certainly, specific endorsement of the project by the host and investing country governments will be necessary in order to ensure that a project meets both national standards and comports with some set of international criteria. Additionally, some international body may be given the authority to review projects.

A strategy which called for national governments to endorse JI projects might require creation or designation of an institution to review projects in accordance with international standards, akin to what the United States has created in its pilot programme, the United States Initiative on JI (USIJI), part of the US Climate Action Plan released in October 1993 (see also Part III, the contribution by Hilton Lee Graham, US State Department). USIJI established an 'Evaluation Panel', comprised of representatives from eight federal agencies, to review and evaluate JI project submissions. The Panel will review and approve submissions within ninety days, and certify net emissions reductions estimated to result from the project. USIJI also sets out criteria - such as whether the project makes adequate provision for tracking - for the Panel to use to determine project acceptability.[5]

Pursuing this type of strategy would require that every national government funds and supports the development of an institution to perform the screening function. Although countries are better positioned to absorb these functions at a national level, from a global perspective, establishing 150 national institutions to screen potential projects could be duplicative and inefficient. Moreover, such an arrangement would still lack independent international oversight, and would force the international regime to rely on individual national judgements.

Despite apparent inefficiencies, national institutions may be better suited to evaluate potential JI projects against national and local development priorities. Perhaps some combination of international and national review will be necessary to satisfy both the international regime and national concerns.

4.7 Guaranteeing project performance

Failure of JI projects can be characterized in two ways:

- nonfulfillment of investor funding commitments; and
- nonfulfillment of emissions reduction pledges.

One strategy for protecting participants from such failures would be for the JI regime to provide some kind of market-wide financial insurance function. The alternative might be to require insurance on a project-by-project basis. Table 8.6 lists some consequences of these alternative scenarios.

---

[5] Despite the inclusion of screening criteria, no specific guidelines yet have been developed to assist the Panel in determining what might constitute 'adequate' tracking provisions. At what level of detail projects will be screened is still to be determined.

Table 8.6. Institutionalized Insurance.

If centralized among one or several institutions		If performed by individual players	
+	-	+	-
- if all participants use same insurer, risk spreading may reduce costs to individual projects	- increases administrative costs of regime - if sole 'agent' proves unreliable, scheme could collapse	- participants decide what level of insurance to carry	- every project must set aside funds to carry full actuarial risk and/or each government would have to bear risk

Although operating a centralized insurance institution would entail administrative costs, costs to individual projects might be lower. If all projects participate with the same insurer, costs could be reduced because risk would be shared, and because individual projects would not need to set aside funds to cover their full risk. An institution would also lower costs by standardizing analytical functions.

Using a third-party insurer is not, however, without risk. The agent ultimately could prove unreliable, for instance by failing to keep adequate reserves or properly evaluate risks.

4.8 Monitoring project achievement

A successful JI regime suggests the need for regular collection of emissions data throughout each project's lifetime. Emissions information will benefit the regime in at least three ways: by satisfying investors that projects are operating properly and that credits will be accorded, by providing timely notification of project failure, and by developing institutional expertise about which projects work and which do not. The international regime could empower either project implementers or independent third parties to gather emissions data or create a new institution to do so. Table 8.7 lists some of the consequences of these alternatives.

Table 8.7. Project Monitoring.

If centralized among one or several institutions		If performed by individual players	
+	-	+	-
- higher degree of accuracy and reliability due to standardization of methodology and equipment - central archiving of records and continuous monitoring could increase institutional learning	- direct and administrative costs of continuous monitoring could be significant	- costs could be low for either continuous or intermittent monitoring	- possible lower credibility for regime as a whole

The value of reliable, standardized information and the institutionalized learning facilitated through central archiving of data might offset the high costs of administering such an institution. And centralizing costs may be more cost efficient than a scenario under which every project develops and finances its own monitoring expertise. Despite possibly significant costs, regular collection of project emissions data by a centralized institution could benefit the regime in terms of credibility and institutional learning. The standardization of data collection and reporting methodology would reduce variability and thus offer reliability of information, facilitating international review. Requiring every project to provide for routine emissions monitoring would also promote learning, by providing timely and accurate information about successes and failures during design and implementation. And greater accuracy derived from normalizing the process would also enhance the value of the information for learning purposes.

Routine monitoring might also be advantageous in disputes over failed projects. In a bilateral regime in which the agreement is essentially a contract between two parties, the JI institution may have no role in dispute settlement other than to deny credit for emissions reductions. But in a multilateral regime, where liability for project failure will affect the overall viability of the JI regime, it will be important to internalize and to formalize lessons from unsuccessful projects. The failed project must be analyzed carefully to determine where in the development and implementation cycle responsibility for failure lies, making it possible to judge similar projects in the future. The costs of continuous monitoring, which in some cases could be quite high, will have to be evaluated against the value of ensuring that environmental goals are met and of increased institutional learning.

4.9 Verifying project characteristics

Verification is expected to satisfy the question of whether reductions in fact occur - the measurability test. Regardless of whether JI unfolds through bi- or multilateral institutions, and whether it includes countries without specific commitments, ensuring the regime's integrity requires near-term institutionalization of some type of emissions verification function.

Several entities could in principle verify emission reductions: project participants, national governments, and independent third parties. Third party entities including national universities; non-governmental organizations; consultants; an international mechanism, either existing or newly-established; or other multilateral institutions familiar with conditions in the host countries.

Information collected by implementers themselves is unlikely to suffice as the sole method for assessing project results and crediting national reduction obligations. Relying on routine monitoring information reported by participants, without conducting any further review, is a scenario which, given the stakes, may not always inspire confidence in the JI regime or protect the atmosphere. Even projects implemented bilaterally among Annex I countries will require verification; in fact, the need may be greatest in these instances because both parties to the agreement may have a shared interest in supplying information that maximizes emission or sequestration effects. While the regime may want to preserve the right of participants to *report* emissions, the regime is unlikely to be perceived as credible if those contributing capital or responsible for project implementation are permitted to *verify* reductions.

One scenario which preserves the right of parties to report reductions independently, yet offers the credibility of independent review and formalizes lessons, would be for project participants to assume responsibility for routine emissions monitoring while having some centralized international institution conduct 'spot checks' over the life of the project. This combination could help assure all parties that promised results were actually achieved before national accounts were credited.

Verification of emissions reductions needs to be considered early, despite expense, complexity and political sensitivities, because it is the foundation of a credible regime. Lessons from the GEF's experience with building monitoring into project design need to be shared, especially the challenge of developing procedures after projects have been designed or implemented.

5 Evolving a Market Architecture

The future evolution of the JI regime is highly uncertain. As one means to achieving their ultimate goal, the Parties to the Climate Convention have chosen a challenging route: the creation of a market for commodities (*i.e.* GHGs) that have never before been traded or sold. To make the situation even more difficult, the commodities are invisible, odourless, tasteless and hard to measure. A number of the operational uncertainties associated with the management of a JI regime have been discussed herein, but many more exist.

The need to deal with all of these uncertainties suggests the appropriateness of a phased approach to evolving the market architecture of the JI regime.[6] For example, it might be prudent to initiate the JI regime with a pilot phase, as discussed at the 9th session of INC that could last two to five years. During this pilot phase, emphasis would be placed on institutional learning and on the examination of a wide array of institutional arrangements. During this phase, no attempt would be made to establish or account for international credits against national commitments for emissions reductions. Nonetheless, individual governments might offer some form of domestic inducements to enterprises which choose to participate in JI projects.

The United States has adopted a parallel approach in conjunction with USIJI. Recent experience with the GEF supports the notion that periodic evaluations of any such regime are essential. A phased approach, employing periodic evaluations, may balance demands to begin JI projects soon against the need to consider fully the equity and efficiency implications of any proposed regime. A phased approach can maximize institutional learning, while promoting innovation and continued experimentation.

The assumption of additional functions by an institution managing the JI regime may offer additional benefits, even while increasing bureaucratic complexity. For example, it could offer new opportunities for organizational learning, concentrating rather than diffusing knowledge about what works in this evolving 'market'. Given that JI is not accepted universally, some level of bureaucracy seems necessary to assure systematic and credible assessment of the regime. The difficulty is in achieving the optimal balance, that is, knowing how much bureaucracy is 'right' for each stage of market evolution. The danger, of course, is that bureaucracies easily become entrenched stakeholders themselves, and become very resistant to reform and change. Unfortunately, there is no simple way to design an optimal market architecture or to predetermine the best institutional form given the variety of expectations of JI and

---

[6] Several Parties to the Convention and commentators have suggested a phased approach. See, for instance, INC, 1993.

the range of potential institutional forms and functions. A step-by-step approach in which all countries learn together therefore appears the practical way to proceed.

References

Newcombe, K. and R. de Lucia, 1993. *Mobilizing Private Capital Against Global Warming: A Business Concept and Policy Issues*, GEF Administrator's Office, February 1993.

INC, 1993. *Intervention by the Representative of Belgium on Behalf of the European Community and its Member States*, INC IX, 19 August 1993.

Environmental Defense Fund, 1993. *Sustainable Development Through Trade in Environmental Commodities*, Washington DC, USA.

Global Environment Coordination Division, 1993. *Proposal for a Venture Capital Fund to Catalyze Private Investment in Greenhouse Gas Mitigation in the Developing Countries*, Environment Department, World Bank Discussion Note, October 1993.

9. Joint Implementation from an International Law Perspective

NICO SCHRIJVER* [1]

1 Introduction

This contribution focuses on JI from an international law perspective. The first section provides some legal data, positions the Climate Change Convention within emerging international environmental law, and deals with the concept and meaning of JI. Section 2 refers to precedents and discusses which international law principles are at stake in an JI arrangement. As section 3 explains, various partners can be involved and a wide variety of legal instruments is available to record JI arrangements. Finally, section 4 addresses the important question of monitoring and enforcement mechanisms.

Positioning the Convention in Emerging International Environmental Law

As of 27 January 1995, the UN Framework Convention on Climate Change (hereafter FCCC or the Convention)[2] had been ratified by 118 States and the EU, a high number two and a half years after the adoption if compared with other multilateral treaties. The Convention was opened for signature during UNCED (June 1992), and entered into force on 21 March 1994, 90 days after 50 States had deposited their ratifications. Among these ratifying States the various groups are well represented, including more than 80 developing States.

The Convention fits the general trend of international environmental treaty-making: systematic regulation at a global level to cope with common problems, cast in general terms, and leaving detailed regulation to be effected at a later stage. The 1979 ECE Convention on Long-Range Transboundary Air Pollution and the 1985 Vienna Convention on the Protection of the Ozone Layer are examples of similar framework treaties. Substantively, the FCCC also fits the modern trend: apart from reaffirming existing principles relevant to international environmental law such as those of territorial sovereignty, resource sovereignty, good neighbourliness, the duty to co-operate and preferential treatment of developing countries, it features newly-emerging

---

* The author is associated with the Institute of Social Studies, The Hague, The Netherlands.

1 This chapter builds on Part I ('Legal and Institutional Aspects') of Kuik et al., 1994.

2 See for an extensive commentary, Bodansky, 1993, pp. 451-588.

C. J. Jepma (ed.), The Feasibility of Joint Implementation, 133–141.

principles and concepts, including the precautionary principle, 'common concern of humankind', intergenerational equity, sustainable development, the 'polluter pays principle', and common but differentiated responsibilities of States Parties.[3]

Meaning and scope of JI under the Convention

Article 4.2(a) of the FCCC provides that certain policies and measures *may* be implemented jointly. Hence, according to the treaty text JI is merely an option. What could it entail? Various interpretations of the term 'joint implementation' are possible, especially a narrow one and a wider one. JI in its ordinary meaning is that two or more parties act together and in agreement for the purpose of carrying out an obligation or a certain task. In the context of the FCCC it has a more specific and narrow meaning, that is that one State Party carries out its commitment partly in the territory of another State in co-operation with that State according to criteria to be decided upon by the CoP. The concept of JI is an innovation which allows industrialized countries to invest in GHG abatement measures outside their own territory where the return on investment would be higher. Of course, the measures must serve to mitigate climate change and should be taken in cooperation with the government of the other country. Taken in a wider sense joint implementation is yet another example of voluntary co-operation or 'joint action' for which precedents abound in modern international law. Various articles of the FCCC provide for forms of international cooperation to serve the purposes of the Convention, for example by cooperation in research and development and by providing access to additional financial resources and technologies in order to enable all countries to achieve sustainable development.[4] However, the term joint implementation as used in Article 4.2(a) does not encompass all such forms of cooperation but should be understood in the narrow sense.[5]

3 See the Rio Declaration on Environment and Development, 1992. For a discussion of this Declaration and principles of international environmental law, see Birnie and Boyle, 1992, pp. 89-127; Koester, 1990, pp.17-8; Sand, 1993, pp. 377-89; and Schrijver, 1995, pp. 126-30 and pp. 227-33.

4 See the penultimate preambular paragraphs, Article 4.1(g) and (h), 4.3, 4.4, .4.5, and 12.3 of the Convention.

5 Kuik et al., 1994, pp. 1-6.

2 Precedents and International Law Principles

Precedents for this concept of JI can be found in forms of international cooperation under: the 1985 Vienna Convention for the Protection of the Ozone Layer; the 1976 Convention Concerning the Protection of the Rhine Against Pollution by Chlorides; and the 1979 ECE Long-Range Transboundary Air Pollution Convention. The 1987 Montreal Protocol to the Ozone Convention provides for the gradual elimination of consumption and production of chlorofluorocarbons (CFCs). The Parties individually undertook specific obligations in this regard, but provision was also made for joint action. The Rhine Chloride treaty provides, among other things, for a substantial reduction of salt discharges into the Rhine, to be gradually achieved in French territory by a jointly funded installation. JI is now also envisaged in The Netherlands - under a 1992 Additional Protocol to the treaty - by way of diversion of salt water from the *Wieringermeerpolder* lake to the North Sea rather than to the IJsselmeer. The latter project, carried out in the Netherlands, will also be jointly financed by the Rhine riparian States.[6] The ECE Convention on Long-Range Transboundary Air Pollution provides for international cooperation for the development of national policies to combat air pollution, especially through mutual information, consultation, research and monitoring. This Convention has been supplemented with a number of protocols, including protocols on the control of emissions of nitrogen oxides or their transboundary fluxes (Sofia, 1988) and the reduction of sulphur emissions or their transboundary fluxes (Helsinki, 1985; Oslo, 1994). The Second Sulphur Protocol provides for JI in the sense of the FCCC.

Lastly, reference may be made to the UN Convention to Combat Desertification, especially in Africa. This Convention, opened for signature in Paris in October 1994, provides that the Parties to it 'shall implement their obligations..., individually or jointly, either through existing or prospective bilateral and multilateral arrangements or combinations thereof'.[7]

While most of these and other examples fall short of the concept of JI as included in the FCCC, they may serve as indicators of new forms of international cooperation, created and intended to contribute to a system in which States undertake to carry out obligations more adequately by enhanced international cooperation.

---

[6] Documents Netherlands House of Parliament, *Tweede Kamer, vergaderjaar 1992-1993, 22 971, nrs. 1-2*, 30 December 1992.

[7] Article 4.1 of the UN Convention to Combat Desertification in Those Countries Experiencing Serious Drought and/or Desertification, Particularly in Africa, 1994.

International law principles at stake

JI raises a number of interesting and sensitive international law principles. They include: territorial sovereignty; State responsibility and liability; common heritage of humankind; and graduation and integration.

As regards *territorial sovereignty*, legally our planet is still split up in almost 200 sovereign States, apart from some international areas such as the high seas, the deep sea-bed and perhaps Antarctica. Sovereignty implies that the government of a State has the power to exercise the functions of a State within a State territory and to arrange its international relations. A corollary of territorial sovereignty is *permanent sovereignty over natural resources* which means that every State has, within the limits of international law, the freedom to dispose of its own natural wealth and resources and to explore and exploit them. This also includes the right to regulate the admission and activities of foreign investors and the right to pursue its own socio-economic and environmental policies. It goes without saying that respect for the sovereignty of the host State should be an important criterion for any JI arrangement. Under the principle of *State responsibility* and *liability* States have a duty to abstain from measures of economic and environmental policy which are incompatible with their international obligations. This principle may raise some difficult problems with respect to JI projects. If a State is in breach of its commitments under Article 4.2 of the Convention to stabilize emissions, and this breach is due to acts or omissions of another State with which it has a JI project, the second State will be presumably be held liable to the first State under the JI arrangement. However, it is not very likely that both States can be held liable for a breach of the commitments under Article 4.2.

So far, the principle of the *common heritage of humankind* is especially relevant in the international law of the sea and outer space law. However, in 1988 the UN General Assembly adopted a resolution on the 'Protection of global climate for present and future generations of mankind'. The Assembly recognized that 'climate change is a common concern of mankind, since climate is an essential condition which sustains life on earth'.[8] Subsequently, this concept of 'common *concern*' rather than 'common *heritage*' of humankind was accepted in the FCCC (first preambular paragraph) as well as in the 1992 Convention on Biological Diversity (third preambular paragraph). Yet, it still emphasizes the common interests and responsibilities which are at stake.

The FCCC distinguishes between various categories of States and identifies common but differentiated responsibilities. The positioning of States is by no means fixed forever. Differential treatment is obviously not meant as a goal in itself, but as

---

[8] UN Doc., 1998, adopted without vote.

a means of recognizing that unequal treaty partners should not be subjected to the same treaty obligations. When economic development and industrialization in the beneficiary State no longer justify differential or preferential treatment, it is only logical to review their status under the FCCC. Precedents for such *graduation and integration* exist in international trade law, for example in the context of the Generalized System of Trade Preferences and the Multi-Fibre Agreement as well as international monetary and financial law and the practice of the IMF and the World Bank. The FCCC provides for a procedure for such a policy of graduation and integration through amending the Annexes.

The applicability of these principles and the actual application and interpretation of the rules of the Convention can easily give rise to disputes between States or a State and a non-State party to a JI arrangement. This emphasizes the importance of a *proper dispute settlement procedure* in any arrangement to be made. Article 14 of the FCCC provides for procedures for the settlement of disputes between States. Apart from the fact that disputes between a State and a company or another non-State entity are not covered, the foreseen procedures seem complex and heavy handed for disputes related to JI. A binding arbitration procedure, along the lines of numerous bilateral investment treaties, is recommendable.[9]

3 Partners in JI and Recording JI Arrangements

The Convention distinguishes different categories of States, each with their own responsibilities. According to the text of the FCCC 'developed country Parties' can pursue JI with three categories of partners:

- other developed countries (so-called Annex II countries);
- countries undergoing a process of transition to a market economy (included in the list of Annex I countries); and
- developing countries (non-Annex countries).

This may lead to differentiating in criteria and in crediting arrangements, if any. The criteria or rules for JI to be established by the CoP should be kept to a minimum and should not be unduly restrictive. They should provide a fair balance and protect the interests of the parties involved. Relevant overall criteria include:

[9] For a discussion, see Peters, 1991.

- environmental effectiveness;
- cost effectiveness;
- equity;
- flexibility;
- practicability; and
- respect for sovereignty of States.

Recording JI arrangements

A variety of legal instruments is available for recording JI arrangements, depending on the objective to be achieved and the sector involved. In general terms, all parties concerned should strive for transparency, efficiency and simplicity. The first phase of a JI project will normally consist of a series of meetings, at various levels, between representatives of the actors involved. It would be wise to make clear arrangements, but at the same time it would also be highly recommendable to avoid unnecessary formalities. Instruments to record the arrangements include:

a. Binding instruments

The following forms of simplified agreements qualify in international relations as treaties:[10]

- *memorandum of understanding* to lay down the points of departure without detailing the arrangements. Basically, it is an agreement to agree;
- *agreed minutes* in order to express the common intentions and understandings;
- *exchange of notes or letters* to lay down the modalities provided they are approved by all concerned;
- *executive or administrative agreements*, for example negotiated between and signed by the ministers of the States involved.

All these forms would not require the cumbersome procedure of ratification or promulgation. Of course, if one prefers to seek maximum legal security, it would be wise to formulate clearcut arrangements in a bilateral or multilateral treaty, for example through a protocol to the FCCC as provided for under Article 17 of the Convention. The great advantage of the latter option is that uniform and binding rules will be established. However, the drawbacks are that: multilateral treaty negotiations

---

[10] According to Article 2.1(a) of the 1969 Vienna Convention on the Law or Treaties, a 'Treaty' means: 'an international agreement concluded between States in written form and governed by international law, whether embodied in a single instrument or in two or more related instruments and whatever its particular designation'.

are normally rather time-consuming; it may raise the danger of a proliferation of instruments (FCCC, protocol, actual JI agreement); and considerable uncertainty may exist as to who will ratify at which moment and who will decline to do so.

b. Non-binding instruments

Agreed guidelines or codes of conduct can under certain circumstances be a relevant alternative. This can be at a bilateral or, in the long run, at a multilateral level, for example by the CoP or by way of a EU or OECD recommendation.

c. Contract

In case non-State entities (company or NGO) are involved, the normal juridical form to record JI would be a contract specifying the jobs to be done.

4 Monitoring and Enforcement

The verifiability and measurability of the effects of JI are crucial. Hence, careful monitoring of projects is necessary for which guidelines and a national self-reporting system could be developed analogous to those of the 1973 Convention on International Trade in Endangered Species (CITES) or human rights monitoring and reporting procedures. At the stage of pilot projects, thorough monitoring, dissemination and publication of the results may increase attention for and interest in JI. In this way an important contribution can be made towards building confidence in JI arrangements. Verification and on-site inspection should be considered, including parallel accounting of domestic and JI efforts. Dispute procedures should be available and easily accessible.

Enforcement of JI agreements may be a problem as States cannot normally be forced to perform against their will and at best monetary compensation can be claimed in case of non-performance. Sanctions are most likely not a very feasible instrument to enforce implementation of a JI agreement.

There are various reasons for not including a sanctions clause, relating to non-achievement of the intended result, in a JI agreement. First of all, it will not be easy to establish the fact that a certain emissions reduction target is not matched due to non-performance of a JI partner. Second, even if this is established, general international law relating to State responsibility and liability will in any case be applicable as noted above. However, at the level of work contracts penalties may play a useful role, for example penalties for delay as normally included in construction contracts. In fact there are few international instruments which do contain sanctions clauses. In the field of international environmental law they include the afore mentioned CITES and the Montreal Protocol.

In conclusion, while monitoring and enforcement mechanisms will play a key role in JI under the FCCC, careful reporting and inspection systems seem for the time being to be the maximum achievable.

References

Birnie, P. and A. Boyle, 1992. *International Law and the Environment*, Clarendon Press, Oxford.
Bodansky, D., 1993. 'The United Nations Framework Convention on Climate Change: A Commentary', *Yale Journal of International Law*, 18, pp. 451-588.
Koester, V., 1990. 'From Stockholm to Brundtland', *Environmental Policy and Law*, 20, pp. 14-9.
Kuik, O., P. Peters and N.J. Schrijver (eds.). *Joint Implementation to Curb Climate Change. Legal and Economic Aspects*, Kluwer Academic Publishers, Dordrecht, 1993.
Netherlands House of Parliament, 1992. *Tweede Kamer, vergaderjaar 1992-1993, 22 971, nrs. 1-2*, 30 December 1992.
Peters, P. 1991. 'Dispute Settlement in Investment Treaties', *Netherlands Yearbook of International Law*, 22, pp. 91-161.
Sand, P.H., 1993. 'International Environmental Law after Rio', *European Journal of International Law*.
Schrijver, N.J., 1995. *Sovereignty over Natural Resources: Balancing Rights and Duties in an Interdependent World*, doctoral thesis, University of Groningen, The Netherlands.
UN Convention to Combat Desertification in Those Countries Experiencing Serious Drought and/or Desertification, Particularly in Africa, 1994. *International Legal Materials*, 33.
UN Doc., 1988. *UN Document A/RES/43/53*, 6 December 1988.

10. Strategies to Head for Joint Implementation

The phased approach vs. bilateral framework agreements between nations

PIER VELLINGA and ROEBIJN HEINTZ (Introduction); and NED HELME and JANET-ANN GILLE* (Sections 2-4)[1]

1 Introduction

(by Pier Vellinga and Roebijn Heintz)

There seems to be general agreement that for the next few years, there will most likely be a 'pilot phase' of JI. This chapter deals with the proper procedure for moving from such a pilot phase into a final phase of full implementation of JI. The question is: what strategy is optimal to lay the foundation for an actual application of JI? In our view two approaches can be distinguished:

- the *phased approach* which envisages three phases of JI - a pilot phase, a dual commitment phase, and a trading phase; and
- the approach of *bilateral framework agreements* in which countries work together to set the criteria and standards for JI projects bilaterally.

In the following we (Vellinga and Heintz) will argue in favour of a phased approach; in the remainder of this chapter (sections 2-4) this point of view will be discussed by Ned Helme and Janet-Ann Gille who will argue in favour of the alternative approach.

The phased approach

We argue that the road towards JI should distinguish three phases: a pilot phase, a dual commitment phase, and tentatively, a trading phase; each phase could be reviewed and adjusted by the CoP.

The pilot phase

Until the CoP has its first meeting in March-April 1995, the INC is charged with the preparation of the activities of the CoP. During the regular INC meetings many of

* Ned Helme (director) and Janet-Ann Gille are associated with the Center for Clean Air Policy, Washington, DC, USA.

[1] In this chapter the contributions by Vellinga-Heintz and Helme-Gille have been put together in order to better represent their different points of view.

C. J. Jepma (ed.), The Feasibility of Joint Implementation, 143–160.

the ambiguous articles of the FCCC have led to intense discussions regarding their interpretation: one of the points of discussion is the meaning of JI itself. According to many, it seems unlikely that JI can play an important role in controlling emissions before 2000. It is argued that developed countries may start focusing on profitable energy efficiency and emission control measures within their own borders. In any case, it is our expectation that for various practical and principle reasons the commitments of industrialized country Parties under FCCC for the year 2000 will be fulfilled by domestic policies and measures alone.

However, many countries have expressed their willingness to start experimenting with JI schemes before the year 2000, arguing that JI pilot projects would clarify some of the (technical) questions and uncertainties surrounding JI. So, the present intermediate period is to be considered a pilot phase to further develop formal rules and procedures. In our view JI projects in the pilot phase should be considered as a voluntary effort, which should be registered along with national reporting requirements in the Convention albeit distinctly. There is no reason why JI projects should not be reviewed and approved by the normal review mechanism of the Convention (in this respect the CoP should develop provisional methodologies for reporting and registration). In our view, the actual emissions reduction achieved in pilot projects after 2000 (but initiated before that year) should be taken into account annually only from the moment on that JI is formally agreed upon. In other words, we would argue that the (total) emission reductions of pilot projects cannot be effected retrospectively.

Development of criteria

As has been pointed out by Spradley in Chapter 7, the FCCC calls upon the CoP to take decisions regarding criteria for JI at its first meeting (Article 4.2(d)). It may be interpreted as the task of the CoP to adopt a set of criteria before the application of JI is officially allowed (the so-called 'top-down' approach). However, we believe that it is better to develop JI and JI criteria through a 'bottom-up' approach. According to the latter approach, the CoP would only make proposals with regard to the process, on the basis of a provisional agenda indicating the direction in which the CoP believes JI should evolve. The provisional criteria, guidelines and methodologies thus provided, could be refined in the course of time, as experience is gained with pilot projects and as the work of the IPCC evolves. 'Learning through trial and error' rather than strict rules is the key. Only in due course, the CoP should aim at adopting efficient and equitable criteria for JI, which take account of environmental, economic and social concerns.[2]

Registration of JI projects

The possibility of a reliable and standardized calculation of the results of a JI project is an essential precondition for a credible and transparent JI mechanism. The IPCC (Working Group I) is currently addressing accountancy issues with regard to emissions from different sources and sinks of GHG. Accounting for JI should conform to the general accountancy principles and methodologies to be decided upon by the CoP. A feasible way of enhancing transparency would be to report about policies, measures and their effects at home and abroad in a national report under separate headings. The host country party could likewise report separately on any effort that has been made with the assistance of others. The results reported through different channels should be examined, reviewed and registered periodically. In our view, in the first stages of a JI mechanism, there would be no need for a large administrative body to monitor transactions. JI, accounted for on a parallel basis, could be reviewed by the normal review mechanism of the Convention.

Dual commitment phase

Once the current commitments under the FCCC would be fulfilled, it may well turn out to become increasingly difficult to establish *identical* commitments applicable to all developed countries. After all, there are basic structural differences (energy use, cost of GHG emission reductions, efforts in the past, per capita emissions, *etc.*) between these countries, which leads to diverging objectives and perspectives of countries.

In negotiating on future commitments (as from CoP 1) this problem could be solved in our view by introducing a *'dual commitment phase'* after the pilot phase once JI is formalized. In this second phase, countries could indicate separately which part of their emissions reductions (in % or in tonnage) would be achieved domestically, and which part abroad. For example, countries may conclude a dual commitment for the year 2010 consisting of a domestic commitment and an additional 'JI commitment'. We anticipate that the domestic efforts of Annex 1 countries will be considerably larger than the efforts abroad.

Dual commitments could be a response to the in our view legitimate concern that the primary responsibility of industrialized countries is to undertake measures at home. An advantage of the 'dual commitment' is that it can avoid international dis-

2 For criteria, see also Kuik et al., 1994.

cussion over credit arrangements because obligations at home and abroad are treated separately. To be more specific, periodically, as part of the FCCC reporting mechanism, the JI commitment would be compared with the actual achievement of this target as indicated in the JI account. However, the question of whether reductions achieved abroad by means of JI projects may be used to offset any non-attainment of the domestic target would crucially depend upon the approval of the CoP.

Trading phase

In order to achieve the ultimate objective of the FCCC, over time, gradually more emphasis should be given to global cost-effective strategies. After one or two rounds of dual commitments, *i.e.* once considerable experience is gained, the international community may decide to put emphasis on global cost-effectiveness. Flexibility is then the key; remaining constraints should be removed. The location of projects to fulfil national commitments should now crucially depend upon cost-effectiveness only, rather than on some type of dual commitment. Investments in JI projects should now be rewarded with real credits, rather than with 'political credits'. In due course, when transactions with JI projects are standardized, these credits could be traded bilaterally or multilaterally.

Incentives to make it work

Incentives at the national level

JI projects are carried out between two or more governments. However, they should encourage the involvement of private companies, because these possess the suitable technologies and the financial resources. Moreover, the business community commonly has gained valuable experience in transborder business relations, and is usually well-informed of the conditions prevailing in potential partner countries. The ultimate responsibility for the JI project, however, would remain with the relevant governments, because they have signed up to the FCCC commitments.

In short, governments should encourage the business community: to cooperate in identifying JI opportunities; to undertake the JI projects approved by the governments; and to invest in JI.

During the pilot phase, governments will have to offer strong incentives to industry to participate voluntarily in JI projects. For example, governments could conclude 'covenants' or issue 'national credits', *i.e.* exemption from taxes or similar measures. Obviously, such national incentives are the primary responsibility of the national governments.

Incentives at the international level

As we anticipate that JI efforts during the pilot phase will involve no international 'credits', other incentives should be offered to induce investments in pilot projects. Within the FCCC, industrialized countries have expressed their willingness to 'take the lead' in tackling the global warming problem, and to 'assist others' in their contribution. Indeed, JI pilot projects could pave the way for global cooperation to address GHG emission limitation. Moreover, countries may invest in JI pilot projects to anticipate a next phase wherein JI actually exists. By now already experimenting with JI, we believe, countries can build confidence with respect to JI, and ensure a genuine spirit of global partnership. It is therefore also important to encourage the dissemination of JI information, to provide opportunities for research on JI, and to encourage the discussion about JI.

As pointed out by Wrexler et al. in Chapter 8, incentives to start experimenting with JI projects could also be provided by supportive intermediate institutions. For example, the Annex I countries could decide to establish a platform for information exchange on potential JI projects to lower administrative costs. This could be done by establishing a register providing information on: the type of projects; their origin; their costs; and their expected abatement effects. In due course, when experience with JI and related credits has been gained, these credits could conceivably be traded multilaterally, *e.g.* via a *'clearing house'* (obviously, a monopoly position to any of such institutions should be avoided).

2 Bilateral Agreements Between Nations

(comment on section 1 by Ned Helme and Janet-Ann Gille)

2.1 The phased approach

In the preceding section it was argued that during the first 'Pilot Phase' all JI projects are voluntary. The INC and, eventually, the CoP, develop *process*-oriented criteria, rather than *project*-directed criteria. The incentive for participation at this stage of the programme is either provided internally by individual nations or wholly for 'social learning' motives. In the second phase nations make dual commitments at which stage the CoP begins the development of provisional criteria that outline how the international commitments will be fulfilled. The incentive for participation at this stage primarily is the fulfilment of each nation's JI portion of total commitment. In the final phase, international trading begins where nations establish their own plans for rewarding or encouraging JI projects and 'credit trading'. Under this phase, the CoP establishes formal criteria for all JI projects.

There are two key problems with this type of phased approach. First, a major incentive for participation in JI projects - recognition for cost-effective reduction in GHG levels to meet a specific target - is postponed to the distant future. In fact, the approach does not envision crediting JI efforts until after 2010 or 2020. This significantly reduces the likelihood of participation by all but national governments interested in the social learning aspects and private industries that are somehow guaranteed acknowledgment within their own country for international efforts.

Second, the phased approach delineates between voluntary projects and projects that involve a return on the investment for the investing parties. These two concepts are not mutually exclusive. Combining no emission reduction or offset acknowledgment with no return on investment, effectively negates any incentive for JI projects to move forward in the next 10-20 years, except where undertaken by national governments.

2.2 Bilateral agreements between nations

There is another, more efficient way, to approach JI. Recently, the Center for Clean Air Policy (the Center) has been working with representatives of two potential JI host countries and the US Department of Energy to establish *'bilateral framework agreements'* that will create a framework for JI projects to move forward. Through these bilateral arrangements, nations work together to set the criteria and standards for JI projects between their countries to ensure effective JI projects and accurate recording of their results. These agreements should be designed generically to serve as a model for eventual development by the INC/CoP of credit trading rules.

At the same time, this approach offers incentives on the national level for JI projects to move forward and provides real-world experiences with JI projects that can serve as a foundation for building the institutional capacity necessary for a full-scale JI programme in the near future.

To establish effective bilateral agreements there are several key features that must be addressed. In developing this concept, the following draft outline of the necessary characteristics for bilateral JI agreements between host and investor nations is proposed:

- Both nations must have national inventories of GHGs and be signatories to the FCCC.

- There should be a tacit agreement allowing for sharing of emissions reductions or offsets.
  - JI projects implemented under the protocol should include a provision that allows for the sharing of emissions reductions or offsets.
  - The ratio of reductions shared between the countries should be determined on a project by project basis.
  - GHG reductions can be transferred between the two countries without being specifically tied to an international investment project. In other words, nations can transfer domestic GHG reductions or offsets to other nations.
  - If reductions are transferred, not tied to a specific project, the country transferring the credits must demonstrate that they represent actual net reductions in GHGs.
- There should be an agreement on compatible monitoring and verification criteria for JI projects.
  - To ensure that reductions or offsets are credible and accurate, nations must demonstrate that a monitoring and verification strategy is in place to measure the net GHG reductions.
  - For specific JI projects, host nations may choose to use the monitoring and verification guidelines established by the investor country, and agree to have the JI investors provide the monitoring and verification.
  - For the transfer of GHG reduction credits which are not tied to a specific project, the country transferring the credits must have an internal monitoring and verification system compatible to the investor's nation.
- There should be agreement to adjust national GHG baselines to accommodate the trading of reductions or offsets.
  - To ensure that emissions reductions are not double-counted internationally, the host country must agree not to include emissions reductions (whether tied to a specific project or as credits) transferred to the investor country in their national emissions baseline. For example, if host country A has an actual 1990 baseline of 1,000 tons and transfers 100 tons of reduction to investor country B, then country A must adjust its 1990 baseline to 900 tons.
  - Investor nations must include the transfer of reductions or offsets in their national baseline.
  - Since the INC does not require reporting of JI results in national annual reports, data on emissions reductions transferred under JI must be kept and recorded by the individual nations.

- There should be an agreement on streamlined government processes to approve projects by types, rather than on an individual basis (*e.g.*, fuel switching or efficiency projects at district heating plants).

Under these types of framework agreements, JI projects will be encouraged through recognition guarantees for participants - both national and private. Also, as opposed to the phased approach, projects that provide a return on investment would be encouraged. Most importantly, the bilateral protocol approach offers the necessary incentives for JI projects to be undertaken now and provides a clear demonstration of their benefits to participating nations.

Concerns have been expressed that through a bilateral protocol strategy, we would end up with over 160 different agreements, each with their own unique features. If bilateral agreements were our ultimate objective, this might very well be true. But they are merely a stepping stone. The ultimate objective for these agreements, as with the phased approach, is movement toward an international trading system based on the models provided through bilateral agreements. However, with the bilateral agreement approach a trading system will evolve more quickly and standards for determining the accuracy and credibility of JI projects will have been demonstrated. In fact, through these agreements we can completely by-pass the need for a Phase Two and move us into a full-scale JI and trading programme after the year 2000.

2.3 The Decin model: a case study

Background

For the past two years, the Center has provided technical and policy assistance to a number of Czech and Slovak communities, in dealing with their severe air pollution problems, from its field offices in Prague and Bratislava. Some areas in this region suffer from such severe air pollution that one-half of the children in local hospitals are suffering from health problems caused or aggravated by air pollution. And the average life expectancy in the region is five years shorter than in areas in the Czech Republic without such vexing air pollution problems.

The USA can - and should - take advantage of every opportunity that will lead to real improvements in this public health and environmental tragedy while also making measurable gains in cutting GHG emissions. The Center has been working with mayors from the Northern Bohemia region of the Czech Republic to alleviate these problems. It has successfully lined up three American electric utility companies - Wisconsin Electric Power Company, Edison Development, and NIPSCo - who will each invest $200,000 in the conversion from lignite coal to natural gas of the Bynov district heating plant in the city of Decin. Energy efficiency improvements to the heat distribution network are also part of the project. Reductions in GHGs will be achieved while also cutting emissions of air pollutants which have so damaged the local environment and shortened the lives of Northern Bohemia residents.

The matching up by the Center of Decin's desire for air pollution reductions and the utilities interest in demonstrating the viability of JI provides an excellent example of the proverbial 'win-win' type of project that can be produced through JI. Given all the other immediate environmental challenges faced by the city, it is quite understandable that the idea of GHG reductions and JI had never crossed the city's 'radar screen'. As a result of this project, the opportunities for more of such projects is appreciated by all of the mayors of the region as well as by the Czech Ministry of Environment. The potential for replicating this first-of-its-kind pure private sector JI project in other Eastern and Central European Communities is enormous. There are an estimated 5,000 coal-fired district heating systems or industrial boilers in the Czech Republic, 15,000 in Poland, 40,000 in Russia, 16,000 in Ukraine, and 3,500 in Hungary. Many are outdated, inefficient, and operate with minimal pollution controls. This case study focuses on the project the Center initiated in the city of Decin.

Decin is a heavily industrialized center with a population of 55,000 located in the northwestern corner of Northern Bohemia in the Czech Republic. The city sits astride the Elbe River in a deep valley about 10 kilometres wide, at the base of 500 meter tall ridges. 80% of Decin's residents live in multi-family flats. Only about one-fifth of the existing housing is in single-family units.

Most of the city's housing units are heated by local brown coal (lignite) - used to fuel the district heating system and individual homes. As of 1992, 75% of the square footage in the community was heated by coal, with about half of this total linked to the district heating system. Because the stacks on the district heating plants are low, emissions from the plants lay within the valley. The extensive use of brown coal is the principal source of the high levels of air pollution - sulfur dioxide (SO_2), nitrogen oxide (NO_x), air toxics and particulates - suffered by the region. While large power plants within the Czech Republic have begun to receive assistance, district heating plants are often too small to attract multilateral funding, despite their devastating effect on the area.

Because of its location in a deep valley and its heavy reliance on brown coal, Decin is one of the most polluted towns in Northern Bohemia. Concentrations of some air pollutants far exceed national and international health standards. For example, the highest twenty-four hour average SO_2 concentration occurred in January 1987 and was more than ten times higher than health standards set by the World Health

Organization. Heavily contaminated air threatens human health throughout the region. For example, childhood mortality rates are twice that of the rest of the Czech Republic. Half of the paediatric patients in area hospitals suffer from conditions linked to air pollution. And, on average, life expectancy for residents in the region is five years shorter than in other parts of the Czech Republic. To prevent the large scale relocation of environmental refugees, the prior regime resorted to financial compensation - termed 'burial money' by recipients - to those who stayed in the region.

Under the leadership of Mayor Kunc, Decin has placed the highest priority on developing specific projects to immediately begin reducing air pollution in order to improve the health of local residents. In the last several years the city has become a leader in the Czech Republic for its efforts in addressing its environmental problems. A year-long environmental study, completed for the city by the Danish consulting firm of Bruun and Sörenson in April 1993, was carried out to provide a series of recommendations to the City of Decin in their attempts to reduce life-threatening pollution. Options and recommendations offered in the study included:

- revamping the district heating system - supply-side efficiency and fuel switching;
- improving energy efficiency in apartment blocks;
- installing energy control equipment in units;
- switching individual houses from coal to gas; and
- switching large industrial users from coal and/or heavy oil to efficient gas systems.

The project

With financial assistance from the three US utilities mentioned above - Wisconsin Electric Power Company, NIPSCo, Edison Development - the Center is working with the City of Decin to develop a first-of-its-kind pilot JI project. One of the city's five district heating plants, the Bynov District Heating Plant, which provides heat and hot water to the multi-family dwellings and several businesses, will be switched from brown coal to natural gas. At the same time the system's efficiency will be significantly improved, allowing the facility to be downsized from 19 to 6 MW.

Currently, the Bynov plant has three 5.6 MW coal-fired boilers and one 2.8 MW coal-fired boiler. By early 1995 the Czech government will extend an existing natural gas pipeline to the gate of the district heating plant. With partial financing from the three US utilities, the city will be able to replace existing coal-fired boilers with new, more energy efficient gas-fired boilers. The project will also replace the existing, poorly insulated, pipes that carry the hot water to the multi-family dwellings.

Presently the plant also loses a great amount of efficiency by converting the hot water to steam at the plant, then recondensing the steam back to hot water at the dwellings. With the new, well insulated pipes, the project will eliminate the need for conversion from hot water to steam and back.

At this time, the apartment blocks as well as the district heating plant are owned by the federal government. It has already been agreed that by summer of 1994, the Bynov District Heating Plant will be privatized under the new 'district heating enterprise' and primary ownership will be relinquished to the city. The apartment blocks will also be privatized under the 'housing enterprise'.

Impacts of the project

The fuel switch and efficiency improvements will reduce air pollution - with benefits to both the regional and global environment. The simple measures being invested in by the three US utilities and the city will have the following impact:

Table 10.1. Impacts of the Project in Decin.

Carbon Dioxide	reduced by 12,800 tons/year *(a reduction of over 65%)*
Sulfur Dioxide	reduced by 96 tons/year *(virtually eliminating SO_2 emissions from the district heating plant)*
Ash	reduced by 3,190 tons/year *(total elimination of this waste stream)*

While the CO_2 reductions will help to reduce the threat of global climate change, the immediate results from the project will be demonstrated most notably in the reduced levels of SO_2, NO_x, air toxics and particulates. For the community of Decin, this project will help the community to continue to improve the poor air quality and poor health that have resulted from uncontrolled emissions of these air pollutants.

Financing the project

In attempting to finance this project, the City of Decin found that to receive a loan, they would have to pay interest rates in the high teens, and pay back the loan, on average, in four to eight years. If they could increase the project down payment the interest rates would decrease and the payback period would increase. With a 40% down payment the city can afford to finance the project. There are two options for Decin within this project: at $1.5 mln. they can convert the boilers from coal to gas-fired boilers and make efficiency improvements on the heat distribution network or, at approximately $7 mln., they can develop this further into a co-generation project. The city has already had discussions with the CEZ - the national electric distribution company - and received their agreement to purchase power from the potential co-generation project. If the co-generation project goes through, it will be a model of up-to-date technology for the region. Also, the larger project, especially with the commitment from CEZ to purchase the co-generated power, is much more attractive to investment.

That's where JI comes in. The three US utilities will provide $600,000 - $200,000 each - to Decin, which in turn will use the capital to finance the remaining portion of the project costs. Costs for the project are estimated to $5.48/ton CO_2 reduced on a total project basis.

Table 10.2. Decin district heating fuel switch (in $; future coal price $30/ton).

annual change in fuel costs	-2,293
annual capital costs	123,371
ash disposal savings	-51,858
permit fee savings	-3,209
maintenance fee savings	-136,314
change in annual costs	-70,303
CO_2 emissions reduced (tons CO_2)	12,819
CO_2 reduction costs ($/ton CO_2)	5.48

3 US Domestic Programmes and JI

What does all of this mean for US companies - like the ones investing in the Decin Project - that are seeking to invest in JI projects? Completion of the international process required to establish the criteria and institutional capacity necessary to produce incentives for JI will be at least one or two years down the road. The international process is time consuming and not likely to yield clear direction even after the first foundation has been established. But, for US companies this does not need to diminish their interest in JI.

3.1 The USIJI and national inventory; voluntary reporting of GHG emissions

The USA has already put in place two domestic programmes that will ensure that efforts of US industry are *'measured, tracked and scored'*.[3] In November 1992 the

[3] US Government, 1993, p. 27.

US Congress passed and, then President Bush signed, the Energy Policy Act of 1992 (EPAct). The law established the first registry of US GHG emissions and, of equal importance, the National Inventory and Voluntary Reporting of GHG Emissions (Section 1605(b)). Less than a year later, President Clinton announced the US Climate Change Action Plan (USCCAP),[4] which includes a US commitment to meeting a goal of stabilizing the national GHG emissions at 1990 levels in 2000 and the US Initiative on JI (USIJI).

The USIJI is designed to encourage US industry to become trail blazers in developing, implementing, and recording their net GHG reduction achievements abroad. The pilot programme was created to garner needed experience and to

> establish an empirical basis for considering approaches to JI internationally and thus help realize the enormous potential for JI both to combat the threat of global warming and to promote sustainable development.[5]

USIJI holds promise for helping to expand the availability of financing and investment in both Annex I and other countries by US companies interested in garnering experience with JI, pursuing some new market opportunities for their products and/or services, and supporting host country environmental and economic priorities.

To register the results of these projects, the USIJI will rely on the above mentioned Section 1605(b) of EPAct. Under this provision, the US Department of Energy is required to establish a programme that allows US industry to voluntarily record their GHG emissions and their efforts to reduce them. The registry is to include measures undertaken both in the USA and in other countries. For example, the USCCAP refers to Section 1605(b) as the mechanism that 'companies participating in the USIJI will use to provide information on the progress of overseas projects...'

The programme is designed to provide US industry with the opportunity to voluntarily record their efforts to reduce emissions at the source, offset emissions through carbon sequestration projects (forestry and agricultural projects) or through off-site activities, such as coalbed or landfill methane recovery projects, either in the USA or internationally. The programme creates a database to record these initiatives and their verified net emissions benefits. This information is expected to be used at a later date - first to track US efforts to meet our stabilization goal and, second, to credit and recognize those forward-looking industries that take the initiative now to reduce GHG emissions and voluntarily record those net reductions under the pro-

---

[4] President Bill Clinton, 1993 Earth Day Address, 21 April 1993.

[5] US Government, 1993, p. 26.

gramme. Among the activities that will be recorded under the programme are USIJI projects. Some US industries have been interested in acting now to reduce emissions, but were concerned that these early efforts would not be recognized. This voluntary recording provision was designed to allay those fears and encourage US industry to be proactive in reducing emissions.

Both of the programmes, USIJI and Section 1605(b) of EPAct, offer US companies sufficient incentive to move ahead with JI projects and expectations that they will be 'rewarded' for their early action. A key problem that arose from the acid rain debates was how to avoid punishing those utilities that had taken measures early to reduce SO_2 emissions. After a lengthy legislative battle, the 'clean' utilities were allocated more allowances than their 'dirtier' competitors through formulas specified in the final bill. The same type of programme can be developed to reward those companies involved in 'early' GHG reduction efforts registered under Section 1605(b).

The development and implementation of the Decin project - and any other international projects which may seek registration under the USCCAP - must be completed with full consideration given to USIJI eligibility criteria. In Part III the USIJI will be discussed in more detail by Hilton Lee Graham (US State Department). His contribution will largely deal with the accomplishment and the contents of the final ground rules and guidelines of this provision.

3.2 Lessons from the acid rain debate

There is concern by some potential US participants in JI (or even other domestic activities to reduce GHG emissions) that, in effect their *'good deed will not go unpunished'*. Some US utilities remember the difficulties that arose, during the Clean Air Act Amendment debates, for utilities that had, voluntarily or by state law, taken action to reduce SO_2 emissions early. Given the uncertainty involved in these projects, why would US companies move ahead?

There are two reasons. First, as a new concept, JI pilot projects must demonstrate the environmental and economic benefits that can be achieved through implementation. Second, the voluntary registry established under Section 1605(b) can provide sufficient incentive for companies to become involved in these projects. The lessons learned during the acid rain debates can be instructive.

Under early versions of the acid rain bill, utilities that had taken early action were getting a much smaller number of allowances than the utilities that had buried their heads in the sand and done nothing. It seemed that the less measures taken, the greater the reward. In the end a provision was added that gave utilities that had significantly reduced emissions prior to the baseline years, 120% of their allowance allocations. At the same time, allowance allocations to the utilities that had not taken early measures were ratcheted down. This ensured that the 8.9 mln. ton cap was unaffected, while allowances were allocated on a more equitable basis.

This exact same process can work for companies that register projects under Section 1605(b) of the EPAct - domestic and international. To ensure that the companies that acted early to reduce or offset emissions of GHGs receive credit, the USA could develop a similar 'GHG allowance' allocation system. For example, if the USA adopted a cap on GHG emissions reducing national emissions to 20% below 1990 levels, those well measured and verified projects that are included in the database will be 'credited' to the project participants. At the same time, the 'GHG allowances' allocated to companies that did not take early action, would have to be reduced by an equal amount. If at that time, the international community still did not accept JI as a component of national GHG reduction plans, ongoing JI projects that are credited to US participants would also have to be compensated for by increasing the emissions reduction burden on non-participating companies.

For example, take a very simple hypothetical national emissions inventory. The inventory includes three companies - A, B and C - all emitting at the same level in 1990 - 100 tons/year. Over the next 10 years, 1990-2000, Companies A and B have growing baseline emissions; Company C's emission level remains constant at 100 tons/year. In 1995, Company A establishes a JI project that transfers 15 tons/year of GHG reductions from the host country to Company A. The project, and the emissions reductions are registered under USIJI and Section 1605(b).

In the year 2000, the country establishes an emissions cap of 240 tons/year (a 20% reduction from 1990 levels) - 80 tons for each company. At this point in time actual emissions for the three companies are as follows: Company A - 110 tons/year; Company B - 110 tons/year; and Company C - 100 tons/year. Because of its JI project, Company A also has 75 tons of offset reductions registered under Section 1605(b). If Company A received credit for them, without any adjustments to the other companies, the emissions cap would be raised to 315 tons - above even a stabilization target.

Instead, to ensure that the cap is maintained, a strategy parallel to the legislative solution on acid rain would be to divide the 75 tons of offsets by the five years in which they were achieved and add that total to the first five years of Company A's baseline (See Table 10.3). Therefore, Company A's 'GHG allowance' allocation is 95 per year for the first five years (2000-2004) instead of 80. At the same time, an equal amount - 15 tons - must be deducted from the emissions baselines of Companies B and C, reducing their 'GHG allowance' allocations by 7.5 tons/year. Their

'GHG allowance' allocation would then be 72.5 per year for the first five years as from 2000; and, total allowance allocations for each year would remain at 240. In addition, Company A will be allowed to continue to offset emissions by 15 tons/year through the ongoing JI project. If JI counts internationally, the reductions from Companies B and C would be unnecessary. Either way, Company A is a winner.

Table 10.3. A 'Greenhouse gas allowance' example.

	Company		
	A	B	C
Emissions (tons/year)			
in 1990	100	100	100
in 2000	110	110	100
JI Project 1995-1999			
5 x 15 tons/year	75	-	-
Emission cap			
in 2000	80	80	80
GHG Allowance/year	80 + 15 = 95	80 - (15:2) = 72,5	80 - (15:2) = 72,5

While we are still years away from a mandatory cap and allowance system, this example helps to explain the rationale for companies who see such controls coming. It also explains why Section 1605(b) is so critical to creating incentives for US participation in JI.

3.3 Recommendations

While the international debate continues, we need to move JI from the drawing board, or the conference table, to actual, on the ground, implementation. While the USIJI starts us in that direction, more can be accomplished through bilateral agreements. No country has industry-specific targets and timetables for reducing or offsetting GHG emissions. Therefore no strong incentive exists for industry to participate in JI projects. A key driver for JI will be the outcome of ongoing debates on the adequacy of commitments under the FCCC. Without tougher mandates there seems to be little incentive for a significant number of JI projects to move forward in the short run. While it is understood that for some countries the primary participant in JI projects will be the national government, at least for the first stages, this is not true for the USA.

Realizing the limited resources available from national governments, the USA is doing all that it can to encourage the participation of private industry. If the JI process is streamlined to facilitate these projects, while maintaining the integrity of the programme, participation is more likely. If, at the same time, we encourage projects that not only achieve GHG reductions, but offer participants a rate of return, more private industry will participate. Bilateral agreements have the potential to move JI projects forward while ensuring that we develop the necessary foundation for a credible, cost-effective international trading system in the not-so-distant future.

4 Conclusions

In this chapter the question was raised what strategies can be used to lay the foundation for a cost-effective and environmentally sound international approach to JI. Two approaches for such a strategy have been discussed. One is the phased approach as outlined in section 1 and the other is a bilateral agreements strategy of the Center for Clean Air Policy outlined in section 2.

Both of these approaches have, as their ultimate objective, the institution of an international GHG emissions trading system; their principal difference lies in the timing of achieving that objective. The bilateral agreement approach compresses the experience with JI projects and international trading of emissions reductions into the initial pilot phase, while the phased approach stretches that experience over 15 - 20 years. Bilateral agreements are not an end in themselves; rather they assist in laying the groundwork for development of an international trading system. An interesting case study of a bilaterally agreed pilot project was provided by the Decin model in the Czech Republic.

From both contributions to this chapter it can be concluded that without industry specific GHG reduction targets in any nation, it is essential to promote JI through incentives such as: social learning; recognition of reduction efforts; encouragement of projects that both bring a return on investment and are climate-friendly; establishment of the necessary international framework for evaluating the effectiveness of JI projects; and contribution of financial resources to assist with transaction costs in the early stages. Promoting these objectives will have a beneficial effect on both the global environment and national economies.

References

Kuik, O., P. Peters, and N. Schrijver, 1994. *Joint Implementation to Curb Climate Change; Legal and Economic Aspects*, Kluwer Academic Publishers, Dordrecht, The Netherlands.
US Government, 1993. *US Climate Action Plan*, October 1993.

PART III

THOUGHTS ON
JOINT IMPLEMENTATION
FROM
THE POLICY PERSPECTIVE

11. Joint Implementation

What the Parties to the Climate Convention should do about it

BERT METZ*

1 Why we need a sound JI system

1.1 Introduction

As stated in the first part of this volume it is the FCCC's objective to stabilize the concentrations of GHG in the atmosphere at a level that would prevent dangerous anthropogenic interference with the climate system. Current scientific assessment points out that the necessary reductions might be in the order of 60% or more compared to present global levels. It is therefore a great challenge to control and reduce global GHG emissions. Although the FCCC states that industrialized countries should stabilize their levels of emission at the current levels, global emissions will continue to grow because of the need of developing countries to alleviate poverty and provide their citizens with decent living conditions. The gains that will be made due to the commitments of industrialized countries to stop the growth of their emissions will in the near future therefore be far outweighed by the growth in emissions in developing countries.

1.2 Potential benefits of JI

Precisely because of the large differences in the costs of reducing GHG emissions between countries and because financial resources are limited, JI offers a unique opportunity to reduce global emissions at much lower costs than through reductions in industrialized countries only. The study by McKinsey (1989) mentioned earlier in this volume[1] indicated a 35% improvement in cost-effectiveness, *i.e.* 35% more emission reduction at the same costs, if a regional approach were replaced by a global one. Compared to a national approach the gains of using a global scope are much bigger. For example, a study by the Environmental Defense Fund for the Tennessee Valley Authority,[2] one of the four largest electricity generating companies

---

* Bert Metz is director of the Climate Change Programme of The Netherlands' Ministry of Housing, Spatial Planning and Environment.

[1] See, for instance Chapter 3 in this volume, section 3.

[2] Environmental Defense Fund, 1993.

C. J. Jepma (ed.), The Feasibility of Joint Implementation, 163–176.

in the USA, showed that a 20% CO_2 reduction in 2010 compared to 1990 levels could be achieved at 10-30% of the costs if the most cost effective world wide regime were applied. In this respect one should mention that in industrialized countries often political costs rather than actual economic costs of action are high. (This is for instance the case when considering abolishing implicit subsidies in road transport or energy prices.)

Through JI industrialized countries can assist developing countries to avoid the traditional path of fossil fuel dependency and unsustainable pollution burden the present industrialized countries are following. JI investments would therefore move the economic development of developing countries in the right direction in order to accomplish economic growth in harmony with the environment, which is the idea of sustainable development. JI should therefore not be limited to forestry only. With respect to the sustainable development objective one could think of JI projects which aim to transfer energy efficient technology to developing countries and which provide for large scale investments in renewable energy sources, low waste processes and urban mass transit systems.

In any case, any successful limitation of the growth of GHG emissions in developing countries will require substantial capital flows from Northern to southern countries. The FCCC has established a financial mechanism, with the Global Environmental Fund (GEF) as its operating entity, which is basically meant to fund the incremental costs of developing countries in taking preventive actions. Although the funding of the GEF will be growing somewhat over the next years, the available amount of money is still quite small compared to the needs which will be at least one order of magnitude bigger.

With respect to the above JI may provide opportunities to raise new funds from public as well as private sources. If current development assistance funding would remain reserved for other than environmental needs - as explicitly advocated by the FCCC - additional financial sources would be a prerequisite for further progress. In this respect the only realistic source of additional finance one can think of would be the private sector. If the proper incentives can be created through JI there is a real possibility that private sector capital flows from Northern to Southern countries will increase in order to support the necessary investments in environmentally friendly technologies.

1.3 Potential risks of JI

JI is not an easy concept. It can be misused and then develop into a counter-productive idea. During the recent sessions of the INC developing countries and environmental organizations mentioned a multitude of potential risks of applying JI, especially when this would include developing countries. The main objections to and potential risks of JI can be classified into three groups:

- equity issues;
- sovereignty issues; and
- the issue of 'leakage'.

Equity

In the FCCC a commitment was established that industrialized countries should take the lead in addressing GHG emissions, since they have contributed by far the most to the increase in GHG concentrations in the atmosphere over the past hundred years and are responsible for about 75% of the current emissions.

However, developing countries fear that JI will allow industrialized countries to 'buy themselves out of the problem' of reducing GHG emissions. JI would furthermore allow them to continue their current production and consumption patterns and to eat more than their fair share of the greenhouse cake. Some people stress that the 'day of reckoning' should not be delayed by allowing JI.

Such feelings, as understandable as they are, should not disturb the carefully drafted balance between the actions of industrialized and developing countries that allowed the FCCC to be signed by so many countries.

Sovereignty

A whole range of potential 'sovereignty pitfalls' can be identified:

1 JI could lead to a potential conflict over land-ownership, especially with tree planting projects. Although such projects currently do not aim at formal land-ownership by investors from industrialized countries, they do lead to long-term contracts over land use, with substantial 'damage provisions' in case the forest would not be maintained. A somewhat similar point is a potential conflict with food production in developing countries, if farmers would be massively attracted by the subsidies provided to grow trees as CO_2 offsets instead of growing food.[3]

2 JI could have a potentially negative effect on the economic development in developing countries. Through JI resources could be taken away from development priorities, by steering the local investment and manpower resources towards donor

3 Parikh, 1993.

driven preferences. JI projects might also lead to pressures from investor countries or investor companies to accept investor owned technology, which may not always be the most appropriate for the host country. Similarly, one should take into account the risk of JI for undermining the domestic manufacturing and technology base due to pressures by the investors to use imported technology. For instance, crediting GHG emission reductions could become a factor in decisions of foreign investors on the location of a new plant, even in situations not specifically designed for climate change purposes. JI could promote competition among investors for the 'best' countries or projects, leaving others 'out in the cold'.

3 Developing countries fear that with regard to emission reduction projects in DCs investors from industrialized countries will 'pick the low hanging fruits' first; investors will only invest in the cheap measures to achieve GHG emissions reductions. Consequently, by the time the developing countries have to invest in emission reduction options themselves, only the more expensive options will be left. This argument of course presupposes future quantitative emission targets for developing just as for industrialized countries, which thus far has not been the subject of serious discussion.

Leakage

Various loopholes have been noted which could result in slower global reductions of GHG emissions, if JI were successfully applied. For instance, JI might reduce or remove incentives for non-Annex I countries to be 'promoted to' the group of Annex 1 countries (with a quantitative emission target) in accordance with Article 4.2(g) of FCCC, because it might turn out to be easier to attract JI capital without having the status of an Annex I country. The net result could be that these countries will continue their growth of emission.

Another leakage issue relates to the so-called double counting of reductions, *i.e.* activities under a JI arrangement are counted that do not exist or are much smaller in reality. At least three varieties of double counting can be distinguished:

1 An investor country claims credits for closing plants/facilities at home under its own national plan and takes a JI credit for setting up a similar plant in another country that is better than what would otherwise have existed in the host country.
2 Host countries inflate their baseline by claiming against a situation which in reality would never have occurred in the absence of the JI project.
3 Host countries inflate their baseline by claiming JI reductions for actions that would have occurred anyway.

Leakage can also result from decisions of industrialized countries to postpone the development of new, more sustainable technology. After all, JI would offer industrialized countries a large reservoir of cheap reduction options that can be reaped with the available technologies. The result could be less innovation and therefore a slower response to the climate change threat in the long term.

In the next sections the question how to set up of a JI pilot phase will be discussed. First, the question of why a pilot phase is needed will be dealt with briefly.

2 The JI pilot phase

2.1 Why we need a pilot phase

Practical hands-on experience with JI is very scarce. This makes a discussion on the role of JI and the criteria that should be applied to ensure a proper application a somewhat theoretical exercise. The lack of experience with JI projects makes it too early to take final decisions on and to formulate final criteria for JI projects at this stage. This would restrict the flexibility which is required for a pilot phase. It is therefore important to learn from the available expertise of those who are already experimenting with specific forms of JI (see for instance the contributions to part IV of this volume, in which several emission reducing projects will be discussed). With the help of practical experiences, methodologies and operational criteria can be further developed as the process proceeds.

Although JI is a politically sensitive topic, there seems to exist considerable agreement, also in developing countries, that a JI pilot phase should start. During such a pilot phase there should be no crediting towards current obligations with respect to emission reductions. Documents on the setting up of a JI pilot phase are currently being prepared by the INC Secretariat to be discussed at the (INC) meetings before the CoP 1 decides on the provisional rules of the game during the experimental phase (see also the time schedule at the end of this chapter). Considering the benefits and risks associated with JI mentioned above, it is a challenge to the FCCC Parties to formulate the pilot phase criteria and rules of the game in such a manner that the 'win-win' potential of JI will be fully utilized.

2.2 How a pilot phase should look

Various issues have an impact on how a JI pilot phase should be implemented. One can think of: the characteristics of the pilot projects that will be permitted under the FCCC umbrella; the necessary criteria for pilot projects; the requirements of the process and documentation of projects; the possible types of projects; and the question whether or not crediting towards current commitments under the FCCC should be permitted.

The pilot phase umbrella: projects, incentives and crediting

At the 9th session of the INC, The Netherlands has emphasized the importance of a formal pilot phase, during which crediting towards current commitments under the FCCC should not be permitted. New pilot projects should be registered with the FCCC secretariat. On the basis of this an evaluation of the projects should later on take place by the Conference of the Parties and its subsidiary bodies based on regular accounting and monitoring of the projects. The number of projects eligible for evaluation and monitoring by the FCCC secretariat and the CoP should not be limited to projects initiated under the FCCC pilot phase umbrella; projects that have already started and which are expected to have positive environmental effects in the long term should also be analyzed and discussed. It should be clear that the pilot projects under the FCCC umbrella should fully be in line with the pilot phase criteria decided upon by the CoP 1.

The pilot phase can also serve as a way to explore the incentives of private sector parties to participate in JI projects. In this respect one could think of the so-called 'first mover advantage' of a commercial enterprise investing in a pilot project in another country. Being a first mover provides this enterprise with considerable early experience with JI projects. The experience with and the position in the JI market gained in this way can place a company in a strategically favourable position by the time a JI regime is officially institutionalized by the CoP.[4]

Another incentive for participating in a JI pilot project could be the public relations gains that can be achieved when enterprises producing, for example, energy intensive goods, decide to invest in environmentally sound technologies. In this way these enterprises can show the public, and their customers, their willingness to contribute to a clean environment in their own as well as in other countries. Companies sometimes even accept statutory obligations to invest in emission reduction

---

[4] An example of first mover advantage are the economic successes achieved by Western companies based on their presence in the Eastern European market before the centrally planned economies moved into the transition period.

technologies or sinks. In general, it can be argued that countries, enterprises and NGOs may feel an incentive to voluntary enter in JI projects, in addition to commitments taken under the FCCC, to show their 'good standing'.

An approach to create JI incentives for commercial enterprises would be to allow them to subtract JI reductions from their tax-base in case an energy/carbon tax is introduced. Such a tax credit should be provided by the national governments. Alternatively, a national government can also provide subsidies to firms in order to stimulate their investments in emission reducing projects in foreign countries. One can, for example, think of favourable depreciation schedules for investments in environmentally friendly projects in the home country or, for that matter, in the host country.

Although The Netherlands' government is of the opinion that crediting for JI investments to meet current reductions commitments should not be allowed during the pilot phase, it supports the idea of future crediting against further commitments under the FCCC. At the 9th session of INC in February 1994 in Geneva, The Netherlands' delegation has made a plea for accepting the possibility that projects which have already started before or will be initiated during the pilot phase should be eligible for future crediting against further reduction commitments. This is especially important if one intends to stimulate the private sector to invest in pilot projects now.

Pilot phase criteria

In order to reap the potential benefits of JI while avoiding its negative consequences a set of criteria should be formulated under which JI would be allowed to operate. In our view the following criteria need to be fulfilled:

Emission reductions should be real

This condition is sometimes called the environmental effectiveness criterion, since it includes aspects such as establishing clear baselines in order to avoid double counting. One must be able to show that a project leads to lower emissions than would otherwise have occurred. As was already mentioned in several earlier contributions in this volume it will be very difficult to come up with credible 'business as usual' scenarios. Suggestions have been made to first look at projects for which baseline questions can be answered easily, such as improving the efficiency of an existing district heating system (the so-called 'fence' approach where projects are separated from the economy as a whole). It remains to be seen, if this approach will be applicable at a larger scale.

It is likely that the baseline condition can more easily be met by Annex I countries that have accepted quantitative obligations under the FCCC. However, even in this group complications may arise if one of the partners (such as an Eastern European country) would face a serious economic recession leading to an autonomous decline of GHG emissions. Such an autonomous decline should obviously not be considered under a JI arrangement.

Some people suggest additional criteria to eliminate double counting. For instance, investor parties should be required to present an overall picture of activities at home and in the host country. Other aspects that can be included are verifiability of emission reductions and clear accounting and reporting arrangements. It might be necessary to require more detailed reporting than is needed now under the FCCC for non-Annex I countries participating in JI projects.

Finally, legal arrangements will be necessary to guarantee the projected reductions to actually come through.[5] The risk that the anticipated reductions are lost due to termination or failure of the JI project requires insurance provisions. Especially in the case of forestry projects a sudden loss due to forest fires is a realistic possibility. But other projects may also fail to deliver the projected reductions. Although in principle actual reductions should be the basis for the credits obtained, some form of insurance will be required to protect against shortfalls.

Additionality

JI arrangements should be strictly separate from and additional to financial obligations of Annex I countries under the FCCC. In other words, JI should never erode the readiness of countries to contribute to the financial mechanism of the FCCC, or to provide development assistance through other channels. The latter aspect may require that Annex I countries can only apply for JI after fulfilling their ODA commitments in terms of a certain percentage of GDP.

As already mentioned above, JI should also be additional to what Annex I countries undertake to meet their obligations under the present Convention. JI reductions should therefore not count against the obligations to reduce emissions to the 1990 levels by the year 2000. This type of additionality is meant to ensure that these countries will indeed reverse the growth of their own emissions.

---

[5] An example of such arrangements are the contracts the FACE Foundation signs with partners for afforestation projects with a 99-year contract period (see also the contribution by Verweij in Part IV).

Approval of participating countries

It is important that a JI project will get approval from the participating countries. In order to prevent the sovereignty problems mentioned in section 1.3 the governments of the countries involved in a JI project should play an important role in the negotiations on the implementation of the project. Credits for emission reductions are always given to Parties to the FCCC, which clearly establishes governments' responsibilities. The private sector partners can therefore not act independently from the national governments if they want their JI projects to qualify for (future) crediting.

To ensure that JI projects contribute to sustainable development and are in line with the economic development priorities of the host country, it is on the one hand important that the host country government can control the implementation of the project. On the other hand it is of utmost importance that JI projects are increasingly focused on areas such as: improving energy efficiency on the supply as well as on the demand side; reductions in methane emissions from gas pipelines; waste treatment; improvement and application of renewable energy options.

To ensure that projects will be socially acceptable, local authorities and non-governmental organizations should be involved in the negotiations on and the implementation of the projects right from the beginning. In a similar vein it is also necessary to involve affected parties in the project during the decision making process. It can therefore be recommended to introduce formal requirements on environmental impact assessment for JI projects.

A special issue is the so-called lifetime of projects. To avoid the host country party to be bound forever to a JI contract, the lifetime of a project should always be limited. In this respect, it might be attractive to consider JI credits as being 'leased' to the investor party for a given period of time.

Requirements on process and documentation

At the 9th session of INC, The Netherlands' delegation has stressed that during the pilot phase strict requirements on the process and documentation of projects are needed. Important elements are:

- Projects that may be eligible for implementation during the JI pilot phase need to be properly documented by the participating Parties and partners.
- In advance of the implementation of a project it is required to collect data on GHG emissions in the absence of the project: the 'base line' documentation. This is a crucial prerequisite for the evaluation of the effectiveness of JI pilot projects;

it should only refer to the project itself and its direct impact on the level of emissions.

Since the project characteristics determine whether the baseline information can easily be documented, one needs agreed methodologies for the assessment of the emission reduction potential of JI pilot projects. Since this is not always easy there is a need to put emphasis on projects with a positive impact. An approach could be to list a limited number of JI investment options: a so-called 'positive list'. Such a list would include projects for which it is relatively easy to directly prove the emission reductions. If a proper baseline cannot be defined, potential investors should only invest in projects on the list, at least in the first instance.

- After a JI pilot project has been initiated, periodical reporting on the progress of the project is required. Reporting should be directed to the governments of the participating countries and needs to include, among others, information on the net emission reductions realized under the project.
- The reports on the projects will be subject to verification by the CoP or a subsidiary body. This verification should include a check whether the pilot phase criteria have been met and whether the reported emission reductions have actually been achieved.
- JI can stimulate projects in developing countries which also have high development priority such as possibly energy efficiency improvement and the reduction of local air pollution. In order to check whether JI projects are indeed in line with the local development priorities the contributions of JI projects to local development should regularly be reported.

Types of pilot projects and the role of governments

During a JI pilot phase various types of projects can be carried out. How projects will look depends on the type of the project, the partners in the project and on the role the national governments play in supervising the projects.

JI pilot projects must, of course, aim to control GHG emissions. Pilot projects can therefore involve energy efficiency improving techniques on the demand as well as on the supply side, provided that their application is in accordance with the national priorities of host countries. The Netherlands' government believes that sink enhancement, and in particular afforestation projects, can equally be in line with the national priorities of developing countries, and should therefore not be excluded from the pilot phase. This is the more so because sink enhancement seems to have the potential to contribute to controlling CO_2 emissions in a relatively cost-effective manner.

The importance of private sector participation in JI pilot projects and the role of national governments has been discussed before. At the one extreme, the national government of an investing country can actively be involved in the formulation, the financing and the implementation of the JI projects. At the other extreme the national government can play a fairly limited role, by only formulating the rules of the game and providing the JI private sector partners with some basic, institutional facilities. The Netherlands' government intends to play a role that is somewhere between these extremes: support pilot projects by providing limited financial resources and play a facilitating role in the implementation of JI pilot projects. It will in general be left to the private sector to initiate JI pilot projects.

Another task that can be fulfilled by the government of an investing country is to carry out JI simulation studies. Through such studies insight can be gained into the emission reduction results of ongoing or recently finished emission reduction projects. Simulation studies can furthermore be helpful in investigating the terms at which credits can be earned by both private sector partners and governments when an official JI regime is established. Finally, governments can stimulate JI by encouraging projects carried out at the level of local authorities. For example, one can think of 'twinning', *i.e.* the implementation of joint projects between cities in different countries, that are closely associated. An other type of activity in this context would be the establishment of a small scale fund in a developing country, jointly managed by participants from the South and the North with the purpose of funding a portfolio of small local JI projects. Obviously, such projects designed at the local level in a developing country have a better chance to be consistent with the local development needs. Governments and NGOs should explore the possibilities for such small scale funds in order to create maximum benefits for developing countries that are supportive to national development priorities.

A managed multilateral fund

One of guaranteeing a balance between industrialized and developing countries in a JI regime could be the establishment of a so-called 'clearing house'. In such a system a multilateral fund will be created in which Parties, in order to become eligible for emission reduction credits, deposit a nominal amount of money per ton CO_2 (the so-called 'intervention' price) in order to achieve GHG certificates. If JI projects are subsequently carried out by the multilateral fund, and the actual costs of a JI project are lower than the nominal amount of money deposited in the fund, a 'dividend' becomes available, which can be used, for instance, for the development of local projects in developing countries. On the board of the multilateral fund the Northern and Southern countries should be equally represented.

The realization of such a managed multilateral fund must of course take place in accordance with the framework and criteria as decided upon by the CoP 1.

The oil producing countries and JI

As long as one does not solve the problems created for countries that are heavily dependent on fossil fuel exports (in particular the OPEC countries), there is little chance of a consensus-based movement quick enough to tackle the greenhouse problem. The question therefore arises: can JI be a way to involve OPEC countries in the FCCC machinery? The answer is yes. One can imagine that a market for 'CO_2 free oil' would be created. Oil producing countries could undertake measures for their clients to compensate for the CO_2 emissions from burning the oil by projects, such as eliminating gas flaring at their oil and gas production sites, investing in renewable energy projects, energy efficiency improvements or in forestry. There would be no need to stop at their own borders, projects in other countries would also be feasible. The consumer would then pay for these compensations through a small fee. This might open up an interesting market for oil producing countries that can also be used to reduce their dependence on oil production.

Figure 11.1. The possible development of a JI mechanism in time.

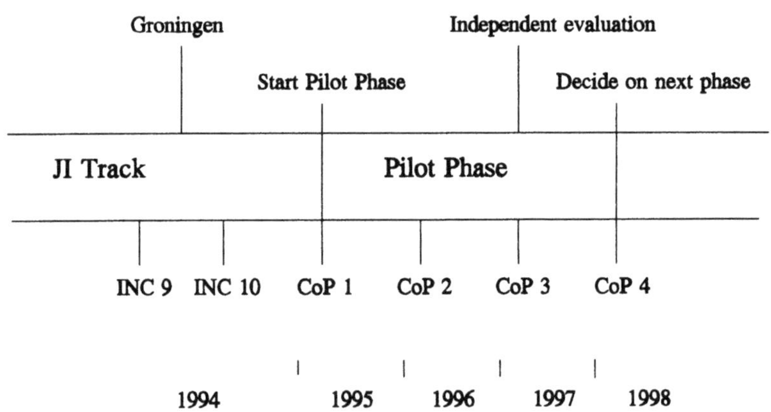

Concluding remarks: the road to CoP 1 and beyond

In Figure 11.1 the possible development of a JI mechanism over time has been pictured. As one can infer from the figure, the Groningen conference has been scheduled as a contribution to a decision on JI by the Parties to the convention. It is envisaged that CoP 1 (28 March through 7 April 1995) will decide on a JI Pilot Phase. Assuming a 3-year pilot period, the pilot phase would come to an end in 1998 after an independent evaluation of the results of the projects. In 1998 CoP 4 could then decide on a new and more mature JI phase.

References

Environmental Defense Fund, 1993. *Joint Implementation: sustainable development through trade in environmental commodities*, New York, August 1993.

McKinsey Cy., 1989. *Protecting the global environment: funding mechanisms*, report prepared for the Ministerial Conference on Atmospheric Pollution and Climate Change, Noordwijk, November 1989.

Parikh, J.K., 1993. *Joint Implementation: a Southern Perspective*, Indira Gandhi Institute of Development Research, Bombay, India.

12. Joint Implementation as a Policy Issue

HILTON LEE GRAHAM*

1 The US View of What the Convention Permits

One of the most discussed - and controversial - parts of the FCCC is the reference in Article 4.2(a) that measures and policies to mitigate GHG emissions may be implemented by Parties jointly with other Parties. The United States believes that the Convention is explicit: per Article 4.2(a), (b), and (d), it fully allows countries to include JI in meeting the 'objective' of the Convention in Article 2.

Because emissions of GHGs everywhere contribute to the build-up of atmospheric concentrations, reductions anywhere can diminish the threat of global climate change. Joint action may also facilitate technology cooperation and a more rapid transfer of environmental technologies from developed to developing countries and to countries with economies in transition. Furthermore, as many measures to reduce GHG emissions have other environmental benefits, JI could stimulate action to reduce multiple current stresses on the environment. In our view, JI can:

- encourage the rapid development and implementation of cooperative, mutually voluntary projects between partners aimed at reducing net emissions of GHGs, particularly projects promoting technology cooperation with and sustainable development in developing countries and countries with economies in transition;
- encourage private sector investment and innovation in the development and dissemination of technologies for reducing net emissions of GHGs;
- provide an incentive, through financial and technical help, for non-Annex I countries to develop their own programmes to combat increases in net GHG emissions; and
- result in cost-effective, world-wide reductions in GHG emissions by encouraging such reductions in the countries and in the types of projects where they are available at least cost.

JI has the potential not only for us all to do more jointly to limit GHG emissions than we could do separately, but also to mobilize significant private sector investment in developing countries and countries with economies in transition. Ultimately, it is through efforts to leverage private sector actions, such as JI, that we

* Hilton Lee Graham is Deputy Director at the Office of Global Change at the US State Department.

C. J. Jepma (ed.), The Feasibility of Joint Implementation, 177–188.

can change the course of the future from its current high GHG emissions path to a lower emissions future consistent with our goals for sustainable development.

2 The Need for Criteria or Ground Rules

We do believe, however, that certain minimum criteria are needed to ensure that JI significantly contributes to achieving the objective of the Climate Convention. To be acceptable, JI projects should meet the following criteria (which largely parallel criteria that the INC Secretariat proposed in document AC.237/49 at the 9th Session of the INC:

- be accepted by the governments of the host and donor countries (approximately equivalent to the Secretariat's possible criterion (S3 for short));
- provide data and methodological information sufficient to estimate current and future net GHG emissions in the absence of, and as the result of, the project (with #3 below, equivalent to S6);
- reduce or sequester GHG emissions beyond those that would occur in the absence of the proposed project (with #2 above, equivalent to S6);
- contain adequate provisions for tracking the actual net GHG emissions resulting from the project, and on a periodic basis, for modifying net GHG emissions reduction estimates and for comparing actual results with those originally projected (no parallel Secretariat criterion);
- contain adequate provisions for external verification of the actual net GHG emissions resulting from the project (no parallel Secretariat criterion);
- identify any associated non-GHG environmental impacts/benefits (approximately equivalent to S5 and S7);[1]
- provide adequate assurance that actual net GHG reduction benefits accumulated over time will not be lost or reversed (equivalent to S8);
- be reported upon by all Parties participating in the project in their national communications, including information on the actual reduction achieved in net GHG emissions and on the share of such reductions attributed to each of the participants (equivalent to S12);
- identify the potential for the project to lead to net changes in GHG emissions elsewhere (no equivalent secretariat criterion); and

---

[1] In this context, we would not endorse requiring a very broad-scoped third party assessment of social and environmental effects. Rather, we would call only for an accounting of the non-environmental impacts associated with the project, in a formulation determined nationally.

- indicate whether efforts are underway within participating countries to develop a national inventory and/or baseline of net GHG emissions, and whether Parties are taking measures to reduce their net domestic emissions of GHGs (no equivalent secretariat criterion).

In addition to these US criteria, the Secretariat proposed several additional points that we could support. These include: S1 (which notes that undertaking JI measures would not 'transfer' commitments to the host country); S2 (which distinguishes JI from obligations to provide financial assistance); S4 (which assumes that JI would not form the totality of a country's efforts, but would be in addition to a programme that would include domestic actions); and S9 (which notes that JI actions could address any GHG).

We would urge that 'operationalizing' the criteria discussed above be left to countries themselves. We also believe that the most efficient mechanism for individual projects to be implemented is the private sector, and recommend that contractual arrangements between actors - consistent with the guidelines promulgated at the national and international level - be the responsibility of the actors involved. We do believe, however, that stringent standards should be met, and that during the course of the review of countries' national communications, JI actions should be closely scrutinized. Based on the review, the Conference of the Parties (CoP) should develop and adopt more detailed technical guidelines.

While finding the majority of the Secretariat criteria appropriate, we cannot accept criterion S10 as proposed in AC.237/49. S10 provides for priority to be given to 'emissions limitations' activities. This implies that sink options would be of a lower rank - a recommendation at odds with allowances in the Convention for actions dealing with both sources and sinks. We do not favour ranking projects by sector or type and instead recommend promoting projects that best meet the above criteria, including acceptability to the host country, most transparent estimation, and credible monitoring of emissions reductions.

In addition, S11, while otherwise acceptable in the permissive form stated, may give rise to requirements that 'credit' be assigned a priori rather than being left to the discretion of the contracting parties. We would not favour such an approach.

3 International Pilot Programme

We believe that the most appropriate way to move toward an international JI regime would be through establishing an international JI pilot programme (see also chapter 13). During this pilot phase, an empirical basis for considering approaches to JI could be developed. The CoP could develop institutional arrangements, detailed technical guidelines, and revized JI criteria and reporting guidelines. Projects in all sectors and with respect to all gases (including both sources and sinks) should be considered, and special efforts should be made to assess how the JI criteria are working in the real world.

A review of the pilot phase could be undertaken at the second or a later meeting of the CoP, at which time amendments to the criteria and methodologies for 'crediting' emissions could be formally agreed.

4 Institutional Issues

In addition to discussing criteria for JI projects, the INC has considered at its session in August/September 1994 institutional issues that would be needed to move forward on JI. Such issues include, but are not limited to:

- *Reporting Procedures.* In our view, JI projects should be included, in a separate section, in countries' national communications. JI project descriptions should include, at a minimum, adequate information to describe how each project meets the above criteria. Formats for reporting on individual gases or sectors should, as much as possible, match those agreed to for reporting on domestic activities.
- *Verification and Monitoring Provisions.* JI projects must contain adequate provisions for tracking and periodic external verification of the actual net GHG emissions resulting from the project. This is particularly important because of the difficulties - especially in non-Annex I countries - of monitoring net GHG emissions. Although the onus should be on the reporting country (which is claiming the emissions reductions in its communication) to assure that the reductions are real, cooperation by the host country will be necessary to verify the reductions properly. We recommend criteria and methodologies for the review of countries' plans to set stringent standards for JI projects, with the burden of proof on the claimant of the reductions.
- *Clearinghouse.* We do not believe that, at this time, a centralized clearinghouse is required for successfully carrying out a pilot JI programme. Rather, collection of information on projects by the Secretariat, compiled based on submissions by Parties in their national communications, will provide a roster of programmes.

5 The International Debate to Date

At the 8th session of the INC in August 1993, a lengthy discussion on JI produced little consensus, largely due to a considerable degree of confusion regarding the provisions of the Convention and possible criteria for JI. In its report on this issue, Working Group I merely agreed to note that the JI discussion had taken place, without addressing any of the (often divisive) substantive issues which had been raised. Countries were invited to submit written comments to assist the Secretariat in the preparation of a new issue paper for the 9th Session of the INC.

Discussion of JI at the 9th Session of the INC in February 1994 marked a major change from 8th Session of the INC. A carefully drawn Secretariat paper served as the basis for discussion, which dealt not only with possible JI criteria but also with the philosophies underlying the various positions. Although differences among governments still exist, and many issues remain to be resolved, the debate was positive, and focused on how to move forward with a workable scheme. Major issues addressed (but requiring considerably more work before agreement may be reached) include:

- *Participation*. While the United States - and many others (see for an overview OECD/IEA, 1994) - support extending participation in JI projects to all countries, others argue that such projects should be confined to Annex I countries.
- *Timing*. Most countries maintain that JI should not be used to meet current commitments to return GHG emissions to 1990 levels by the year 2000; the Convention contains no such restriction, however. Several delegations noted that, in the absence of formal commitments in the convention to limit emissions, this issue is largely moot.
- *Pilot phase*. Nearly all countries (including the United States) supported the Secretariat's proposal for beginning with a pilot programme; this would allow countries to experiment with specific criteria, and see how they measure up in the real world under operational conditions with test projects. The INC asked that in its preparations for the 10th Session of the INC the Secretariat elaborate further on possible goals and opportunities for a pilot programme.

The 9th Session of the INC marked the beginning of agreement on many individual criteria for JI - such as on the need for host country involvement and support, for monitoring emissions reductions, and to assure that projects produce credible and verifiable reductions. That development and the moderate tone of the discussion suggested that the INC should be able to continue substantive debate at its 10th session.[2] At the 9th Session of the INC, the Secretariat was asked to develop options for a phased approach to JI beginning with a pilot phase, and to draft a set of possible JI criteria and institutional arrangements for a pilot programme. Delegations were invited to submit comments on JI to the Secretariat; the Secretariat would use these submissions in developing its options paper for 10th Session of the INC.

It is noteworthy that the text of the Convention (Article 4.2(d)) states that *'The Conference of the Parties, at its first session, shall also take decisions regarding criteria for joint implementation;'* it does not state that the Parties *'shall adopt'* criteria. Thus a strict reading of the text could be construed to allow Parties to defer formal decisions beyond April 1995. The United States would hope, however, that the first CoP will be able to agree on the modalities of a pilot programme to test JI on an international scale.

6 The US Initiative on Joint Implementation

A domestic pilot programme, the US Initiative on Joint Implementation (USIJI), is an important component of the US Climate Change Action Plan announced by President Clinton in October 1993, and was presented to 9th Session of the INC. The Plan's goal is to meet (the US) national commitment to return US emissions of GHGs to their 1990 levels by the year 2000. It will do so by initiating some 50 specific measures and programmes within the United States. The plan is an important first step in dealing with the threat of global climate change.

At the same time, the plan recognizes the potential for significant, cost-effective reductions of GHGs through cooperative projects with other countries. The USIJI seeks to promote a broad range of cooperative, mutually voluntary projects to test and evaluate methodologies for measuring, tracking and verifying costs and benefits. The programme will evaluate project proposals based on criteria designed to promote real and verifiable emissions reductions.

[2] This contribution has been written before INC 10 was held.

7 Development of Final Ground Rules

These criteria or ground rules have been developed through a public, participatory process, including a White House-sponsored Conference on Global Climate Change in June 1993 involving over 800 participants, and a public workshop in July 1993 specifically addressing JI. Draft criteria, published in the *Federal Register* in December 1993, were open for public comment. Twelve different organizations made comments on the ground rules themselves and on the implementation process. The final ground rules incorporate many of the suggestions received.

Some examples of the way in which these comments were or were not incorporated in the final ground rules may be of interest to those who may in the future go through a similar national process. We revized the language under *Purposes*, for example, to maintain appropriate parallelism with the Convention. Under *Evaluation and Reassessment*, we left to the discretion of an Evaluation Panel we are creating whether to open its deliberations to the public. Under *Eligible Participants*, we extended participation to 'groups' of entities, but turned down suggestions to more stringently restrict the eligibility of foreign participants.

We required the *Evaluation Panel* to develop operational modalities, providing it with an opportunity to assure public participation. We rejected arguments that the Panel should accept all projects unless they were clearly deficient, leaving the burden on the proposer to demonstrate adequacy. The Panel has been given the power to oversee the development of the day-to-day operations of the USIJI. We rejected a proposed 30-day turnaround for project approvals/rejections as providing too little time for adequate review of submissions.

Under *Criteria*, we maintained a more stringent 'must find' requirement setting a credible minimum standard. We also defended the essential need for host government acceptance - although how that is to be documented has yet to be defined. We noted that we have to examine how other countries' governments treat JI projects during the pilot phase in order to develop an empirical database for developing appropriate criteria in the operational phase.

While leaving to the Evaluation Panel the development of modalities to ensure that projects involve real, measurable, actions, we also added language to ensure 'additionality:' the measures to be taken are above and beyond those that would have been taken otherwise. We also added a requirement that submissions include data and methodological information sufficient to establish a baseline of current and future emissions - both in the absence of, and as the result of, the specific measures to be taken. The Panel was given authority to develop operational modalities for monitoring implementation of projects, and retains discretion to evaluate possibly negative side-effects.

A criterion requiring registration of the project in the national inventory established under section 1605 of the Energy Policy Act of 1992 was deleted because guidelines for this registry have not been completed; once that is done, however, projects may expect to be registered under the 1605 programme. We decided to leave decisions on apportioning 'credit' for projects to the participants. Despite some concerns about redundancy, we decided to retain Section V.B., which provides additional criteria that the Panel is to consider in deciding whether to include a project submission.

After receiving a number of questions about usage of the word 'net' throughout the text, the words 'reduced or sequestered' have replaced 'net'. We interpret 'reduced' also to include 'avoided'.

The final ground rules were published in the US *Federal Register* on 2 June 1994. In the US private sector, pilot efforts are already underway to develop projects for JI; initial proposals were expected to be submitted within the following several months.

8 USIJI Objectives

The objectives of the US pilot programme are to:

- develop international cooperative efforts towards reducing GHG emissions which contribute to global warming;
- establish an empirical basis for appropriate projects which can successfully combat the effects of global warming and promote sustainable development; and
- achieve greater levels of emission reductions which surpass the capabilities of individual national plans.

The approach is to:

- foster pilot projects among countries and groups interested in reduction and mitigation of emissions; and
- evaluate successful projects based on established criteria for inclusion in the US National Communication to the Framework Convention on Climate Change.

9 USIJI Final Ground Rules

Section I - Purpose:
to:

- encourage the rapid development and implementation of cooperative, mutually voluntary, cost-effective projects between USA and foreign partners aimed at reducing or sequestering emissions, particularly projects promoting technology cooperation with and sustainable development in developing countries and countries with economies in transition;
- promote a broad range of cooperative, mutually voluntary projects to test and evaluate methodologies for measuring, tracking and verifying costs and benefits;
- establish an empirical basis to contribute to the formulation of international criteria for JI;
- encourage private sector investment and innovation in the development and dissemination of technologies for reducing or sequestering emissions of GHGs; and
- encourage participating countries to adopt more complete climate action programmes, including national inventories, baselines, policies and measures, and appropriate specific commitments.

Section II - Evaluation and Reassessment:
The pilot programme will be evaluated and reassessed within two years of its inception or within six months of adoption of international criteria for JI by the CoP to the Climate Change Convention, whichever is earlier.

Section III - Eligible Participants:
- Domestic: Any US citizen or resident alien; any company, organization or entity incorporated under or recognized by the laws of the United States or group thereof; or any US federal, state or local government entity.
- Foreign: Any country that has signed, ratified or acceded to the Climate Change Convention; any citizen or resident alien thereof; any company, organization or entity incorporated under or recognized by the laws of such countries, or group thereof; or any national, provincial, state, or local government entity of such a country.

Section IV - Evaluation Panel:

Co-chaired by representatives of the Department of Energy and Environmental Protection Agency, the Evaluation Panel will include representatives of six other agencies - the Agency for International Development and the Departments of Agriculture, Commerce, the Interior, State, and Treasury.

The Panel will be responsible for: advising and assisting prospective US and foreign participants on technical parameters; accepting project submissions; reviewing and evaluating submissions, including baseline projections; approving or rejecting submissions, based on approved criteria; making publicly available written reasons for its decisions within 90 days; certifying emissions reduced or sequestered estimated to result from projects; developing operational modalities for implementing the Programme; and preparing an annual report of its activities.

Section V - Criteria:

The Evaluation Panel must find that a project submission:

- is acceptable to the government of the host country;
- involves specific measures to reduce or sequester emissions initiated as the result of the USIJI, or in reasonable anticipation thereof;
- provides data and methodological information sufficient to establish a baseline of current and future emissions:
  - in the absence of the specific measures proposed; and
  - as the result of the specific measures proposed;
- will reduce or sequester emissions beyond those referred to in (3)(a) above, and if federally funded, is or will be undertaken with funds in excess of those available for such activities in fiscal year 1993;
- contains adequate provisions for tracking the emissions reduced or sequestered resulting from the project, and on a periodic basis, for modifying such estimates and for comparing actual results with those originally projected;
- contains adequate provisions for external verification of the emissions reduced or sequestered by the project;
- identifies any associated non-GHG environmental impacts/benefits;
- provides adequate assurance that emissions reduced or sequestered over time will not be lost or reversed; and
- provides for annual reports to the Evaluation Panel on emissions reduced or sequestered, and on the share of such emissions attributed to each of the participants, domestic and foreign, pursuant to the terms of voluntary agreements among project participants.

In determining whether to include projects under the USIJI, the Evaluation Panel will also consider:

- the potential for the project to lead to changes in emissions elsewhere;
- the potential positive and negative effects of the project apart from its effect on emissions reduced or sequestered;
- whether the US participants are emitters of GHGs within the United States and, if so, whether they are taking measures to reduce or sequester such emissions; and
- whether efforts are underway within the host country to ratify or accede to the Climate Change Convention, to develop a national inventory and/or baseline of emissions by sources and removals by sinks, and whether the host country is taking measures to reduce its emissions and enhance its sinks and reservoirs.

10 Conclusion

I trust that this exposition of the views of the United States on an international regime for JI, and explanation of the domestic pilot programme that we are initiating to 'test the waters', has been of interest. We hope that it will further the discussion of and debate about how we all can best work together in the effort to reduce the threat of global climate change.

References

INC, 1994. *Document AC.237/49*, 9th Session of the INC, February 1994, Geneva, Switzerland.
OECD/IEA, 1994. *Climate Change Policy Initiatives, 1994 Update*, Volume I, OECD Countries, Paris.

13. Energy Policies in Some Developing Countries and Joint Implementation in the FCCC

WAHJUDI WISAKSONO*

1 Introduction

The availability of sufficient amounts of energy is, next to financial investments, one of the many developmental concerns of Developing Countries (DCs). Although the DCs are fortunately endowed with sufficient indigenous energy resources, they must pay careful attention to consumption, production and availability of energy. As stated by the ASEAN Heads of Government, in their 1992 Singapore Declaration, this especially applies for the six South-East Asian countries (ASEAN countries[1]) in the economically active Asia-Pacific region:

> ASEAN recognizes that sustained economic growth requires considerable inputs of energy. As member states continue to industrialize and strengthen their industrial base, ASEAN shall focus and strengthen cooperation in energy security, conservation and the search for alternative fuels. (Singapore Declaration of the 1992 ASEAN Heads of Governments Meeting; 27-28 January 1992.)

The ASEAN Countries are each in their own way involved in development programmes which are expected to continue well into the next century. As one can infer from Annex 1 the ASEAN countries are consuming energy in progressive quantities. Using Energy Toolbox (ETB),[2] an energy planning software system, the Asian-EC Energy Management Training and Research Centre (AEEMTRC) has forecasted that

---

* Wahjudi Wisaksono is director of the ASEAN-EC Energy Management Training and Research Centre (AEEMTRC) in Jakarta, Indonesia and professor at the faculty of Landscape Architecture and Environmental Technology at Trisakti University, Jl. Kyai Tapa, Jakarta, Indonesia.

[1] The ASEAN countries are: Thailand, Philippines, Malaysia, Brunei Darussalam, Singapore, and Indonesia.

[2] ETB is an integrated energy planning software system with a common forecast methodology developed for the six ASEAN countries. The scenarios used by ETB for ASEAN 2010 were chosen to indicate their scope of *policy options* rather than to display sensitivity to economic growth.
- in the 'drift' scenario energy demand will be completely determined by market forces;
- the 'efficient' scenario broadly foresees maximum utilisation of market forces to promote energy efficiency; and
- the 'tough' scenario includes options which require major policy induced shifts.

C. J. Jepma (ed.), The Feasibility of Joint Implementation, 189–208.

by the year 2010 the final energy demand in the six ASEAN countries will be almost three and a half times higher than in 1988 (see Figure 13.1).[3]

Since the late 1960s and the early 1970s the DCs recognized the existence of a link between development and environment as reflected, for instance, in the 1972 Stockholm Conference.

Figure 13.1. ASEAN forecasted final energy demand (1988-2010).

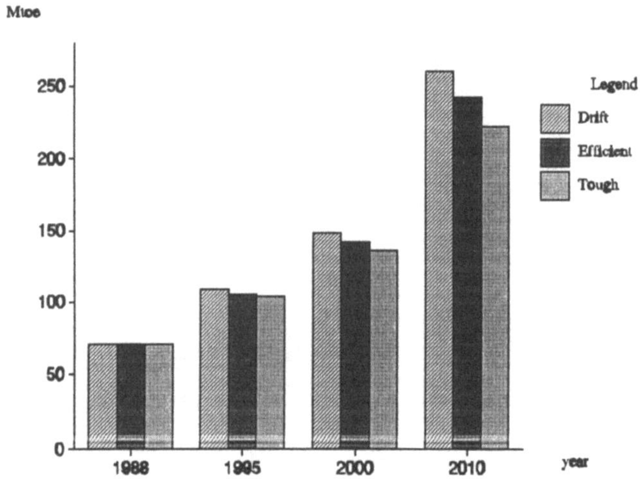

The ASEAN countries adhere to the principle of sustainable development. This is also reflected in the various policies supporting and guiding the development process. The ASEAN countries' policies concerning energy are laid down in *The Energy Policies*, consisting of policies which should satisfy the short term as well as the longer term energy requirements of the countries, taking into consideration quantities, intensities, the environment and other factors supporting sustainable development.

2 Energy Policies in ASEAN Countries

In the Tables 13.1A-F of Annex 2 the policy objectives and policy measures of the six ASEAN countries are summarized, which provide guidance for energy development planning and energy utilization in these countries.[4] The policy objectives and

---

[3] AEEMTRC, 1993a.

[4] AEEMTRC, 1993b.

measures stipulate, among other things, that sufficient amounts of energy, such as oil, gas, coal, geothermal and biomass should be available.

As shown in Figure 13.1, the forecasted increase in ASEAN final energy demand will be considerable in the next decades. Hence, ways and means should be found to utilize energy more efficiently. The 'efficient' and 'tough' scenarios in Figure 13.1 show that energy saving could be obtained through correct and timely policy measures. With respect to these policy measures one can think of activities such as energy efficiency improvements in buildings, transportation, electricity generation and transmission, and in industry. Another option to improve energy efficiency can be through Demand Side Management (DSM), which is currently being promoted in the ASEAN region. Finally, new power plants should use combined cycle or cogeneration principles and processes.

In accordance with the concept of sustainable development, the energy policies do not neglect the environmental dimension. In the Indonesian case, for instance, the reduction of CO_2 emissions, in its energy policy to be achieved through fuel substitution and enhancing energy efficiency measures, is even mentioned.

Figure 13.2. Final energy demand for all sectors in Indonesia.

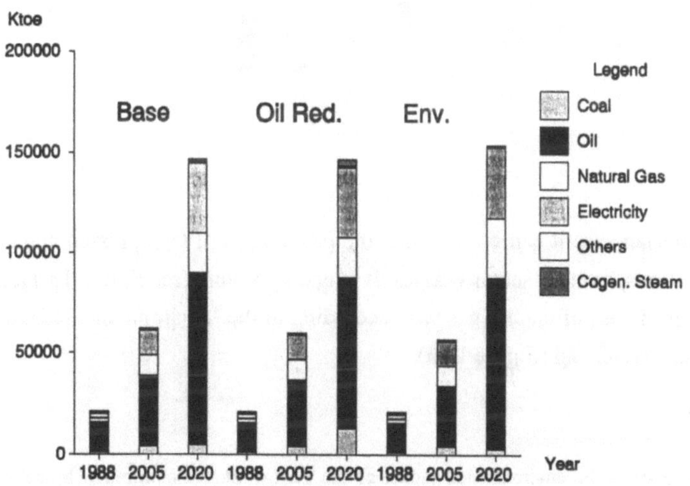

The diversification of available energy sources is depicted in Figure 13.2, which reflects the Indonesian case. This shows a tendency to move away from oil. Computer simulations show that when one intends to minimize the use of oil, natural gas, coal and other sources (such as geothermal and nuclear energy) will become more important in the energy mix. Using the EFOM-ENV model[5] to simulate the diversification of energy use in 2020 natural gas stands out as an important fuel when looking at the fuel with the lowest CO_2 emissions (see Figure 13.2 and 13.3).[6] In South-East Asia the domestic use of this relatively environmentally sound energy source is expected to increase in the coming years.[7]

Figure 13.3. Final energy demand for industry in Indonesia.

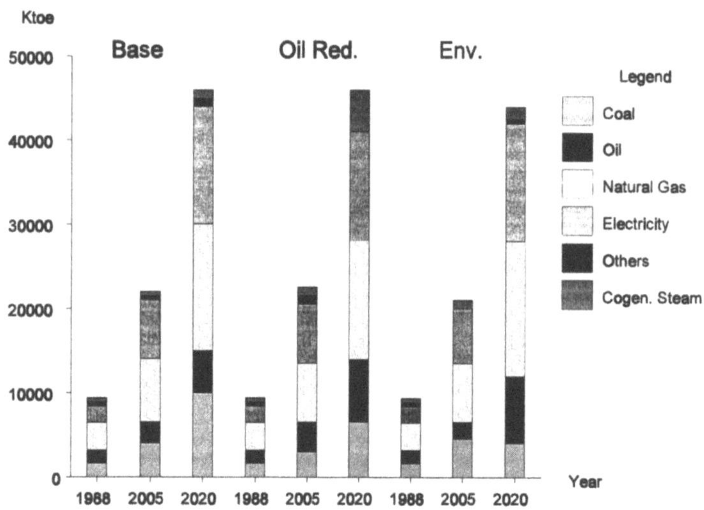

For the Indonesian case it can be seen that the penetration of Compressed Natural Gas (CNG) in the transportation sector will result in energy saving (see Figure 13.4). For the industry sector the use of natural gas will, according to the 'environment' scenario, also increase at the cost of coal (Figure 13.3).

---

5 EFOM-ENV is the environmental model of the Energy Flow Optimization Model (EFOM), which is used in the EU.

6 The scenarios that have been distinguished in the EFOM-ENV model to simulate the diversification of energy use in 2020 are:

- the 'base' scenarios, which have the same underlying assumptions as the above mentioned 'drift' scenario;
- the 'oil reduction' scenarios aimed at policies to reduce oil utilisation; and
- the 'environmental' scenarios which aim to achieve an environmental sustainable growth of energy production and use selected cost-effective SO_2, NO_x, CO_2 control measures and policies.

7 Wisaksono, 1993.

In the household and commercial sectors the use of natural gas, oil and electricity will increase at the cost of coal (see the 'base' and 'environment' scenarios in Figure 13.5). In these sectors an increase in the use of coal is only foreseen in the oil reduction scenario (Figure 13.5).

An important trend in the energy policies in the ASEAN region is the promotion of private participation and private investment in the energy industry, especially in the

Figure 13.4. Final energy demand for transportation in Indonesia.

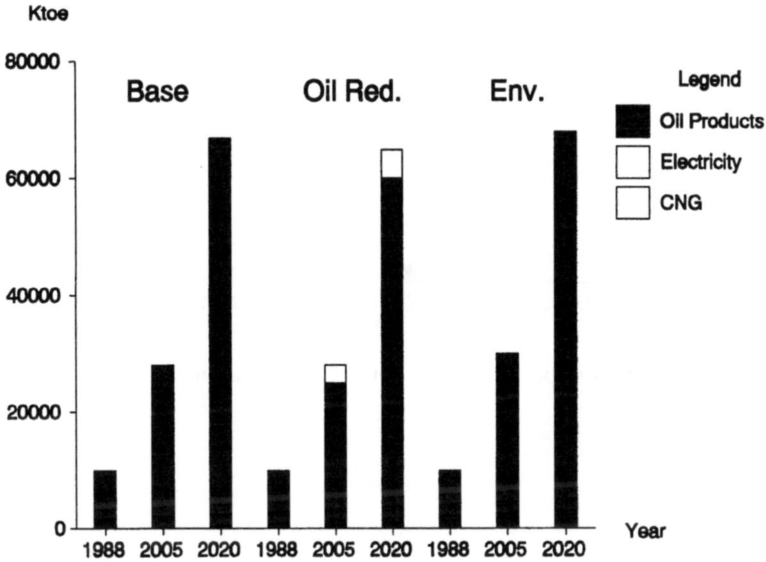

power generating industry. The region's capital investment needs at the beginning of the next century are estimated by AEEMTRC at $378 bln..[8] In this estimate environmental costs are not even included.

---

8 AEEMTRC, 1993a.

Figure 13.5. Final energy demand for tertiary plus domestic sector in Indonesia.

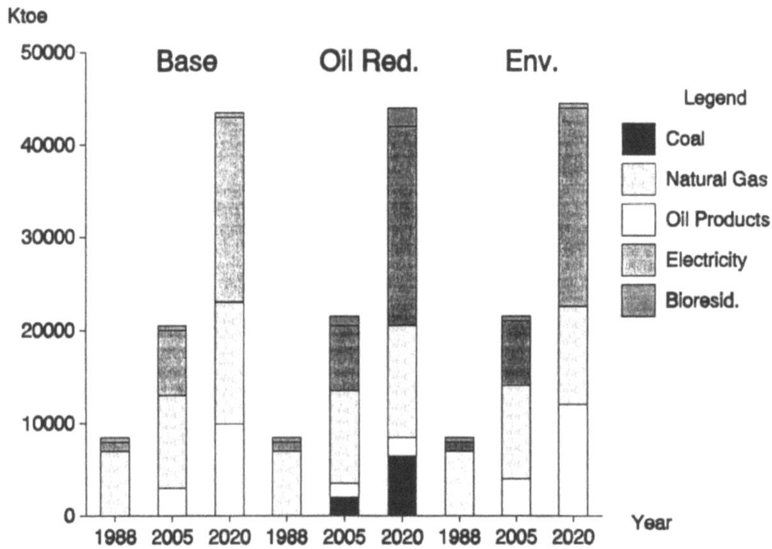

Figure 13.6. Reduction percentage of total CO_2 emissions in Base & Oil Red. scenario.

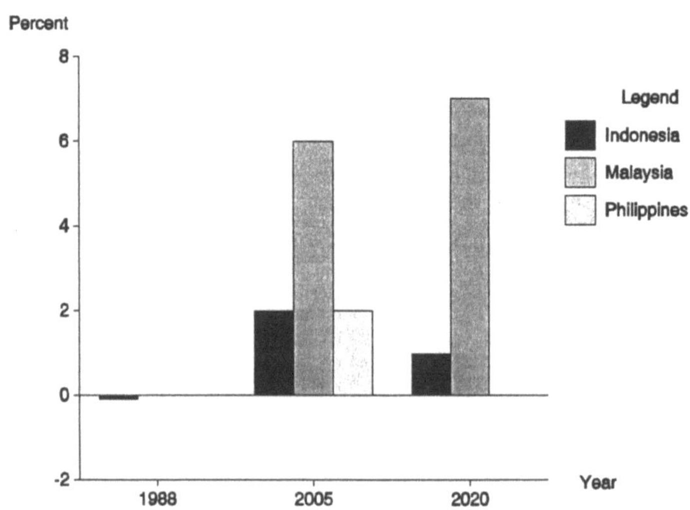

3 Environmental Issues Related to Energy

The environmental issues and energy related problems in the ASEAN region are many and diverse of nature. They can be *local*, such as oil spills and coal ash disposal, or *regional* and *transboundary*, such as the impact of SO_2 and NO_x emissions from industry with the possibility of transnational acid precipitation. The environmental issues can furthermore be of *global* nature, such as GHG emissions which may cause climatic changes in the long term.

The control of CO_2 emissions should therefore be implemented globally and should preferably be cost-effective. For the DCs the implementation of CO_2 emissions control should for obvious reasons be on a 'no-regrets' basis.

As was pointed out earlier the energy policies in the ASEAN region have already resulted in programmes and activities in this direction. In Figure 13.1, for instance, it was shown that energy policies through the AEEMTRC 'efficient' and 'tough' scenarios can exert a downward influence on final energy demand. Furthermore, as can be inferred from Figure 13.6, through comparing the 'oil reduction' scenario with the 'base' scenario, the percentage reduction of CO_2 emissions through a fuel switch away from oil and coal amounts to 2% in Indonesia and the Philippines and 6% in Malaysia in 2005, and to 1% in Indonesia and 7% in Malaysia in 2020. Saswinadi et al.[9] found that price deregulation can strongly motivate such a fuel mix.

Another set of programmes to downwardly influence the CO_2 emissions are the cost-effective energy efficiency enhancement programmes, which are stipulated in the *Energy Policies*.[10] Figure 13.7 shows that in the year 2020 the 'environment' scenario can, in comparison with the 'base' scenario, result in CO_2 emissions reductions of 37% in Indonesia, 18% in Malaysia and 42% in the Philippines.[11]

The control of CO_2 emissions related to fuel-combustion requires international cooperation, which would benefit the global stabilization and reduction of CO_2 emissions. This could involve more intensive regional cooperation on these programmes and more intensive cooperation among DCs and between DCs and industrialized countries (ICs).[12] Within the framework of these cooperation programmes,

---

[9] Sasmojo et al., 1991.

[10] In a study on CO_2 emissions in The Netherlands, Blok et al. (1993) calculated that 41% of the reduction can be achieved through retrofitting, process optimization, etc.

[11] These figures are only for purposes of comparison to indicate changes.

[12] AEEMTRC, 1993a.

mechanisms, procedures and criteria for technology transfers should be developed. Such programmes to control CO_2 emissions should be regarded as programmes to adjust DCs' economies to global CO_2 constraints, while continuing national and regional development programmes, so that the goods and services required in the DCs' national development programmes can be further produced. In the next section attention will be paid to Joint Implementation as an instrument to facilitate such proposed cooperation programmes.

Figure 13.7. Reduction percentage of total CO_2 emissions on Base & Environment scenario.

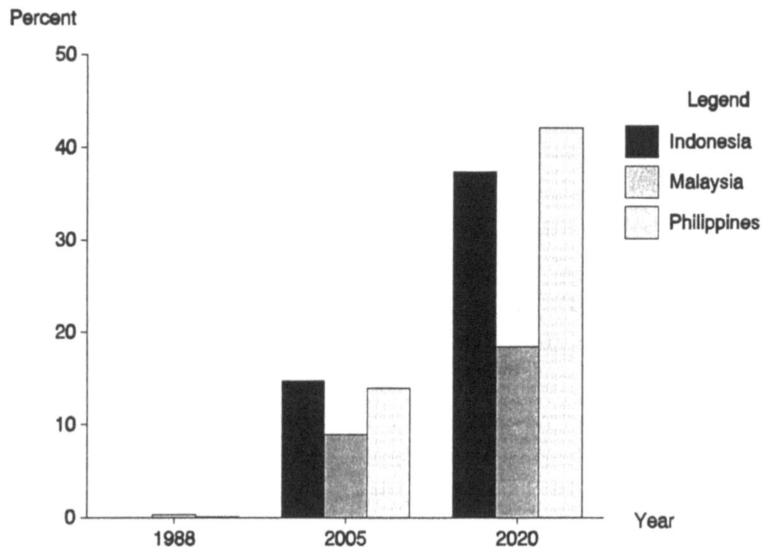

4 Joint Implementation, Factors and Issues to Be Taken into Account

The JI option has been defined and described in many ways, such as:

- a range of cooperative arrangements between countries, designed to reduce the risks of rapid climate change in a cost-effective manner; and[13]

[13] Mintzer, 1994.

- the realization of emission reductions by one party (investor) on the territory of another (host) country.[14]

All Joint Implementation (JI) projects, however, share the characteristics of linking the release of GHG in an industrial country with an offset that reduces or absorbs a comparable amount in another country. This can be a DC or otherwise. As mentioned earlier in this volume, in neither case are the terms and conditions for JI clearly stated.

It is also not explicitly stated whether or not DCs shall participate in JI.[15] The question that arises is how to participate when there is a proposal from a DC or an IC to set up a JI project. It seems that there are at least two options available for DCs:

- to implement CO_2 emission reduction jointly with an IC or IC-party; and
- to receive assistance from an IC or IC-party,

in order to achieve a CO_2 emission reduction or to participate in stabilizing CO_2 emissions in the year 2000 at 1990 levels.

These options have a different character. In the first option JI is viewed as a partnership based on inputs from both sides: the DC brings in environmental capital in the form of unused CO_2 capacities, whereas the IC brings in capital investments and/or technology. Net CO_2 reductions thus obtained should be divided between the two parties. This raises other questions and issues, such as:

- How to value unused CO_2 capacity in monetary terms?
- What formula should be used to divide the net CO_2 reductions?

In case of the second option, the relation between the IC and the DC is a donor-receiver relationship rather than a partnership.

Both options, however, could be integrated into one programme. The programme could start with training courses for which financing might be obtained from the GEF or bilateral assistance. These courses might then be followed by a JI project. Both options can be used to enhance energy efficiency programmes or to introduce technologies that release little or no GHGs, *i.e.* programmes that support the earlier mentioned energy policies.

---

[14] Moomaw, 1994.

[15] Metz, 1994.

Other principles concerning JI that need serious and careful attention are the following:

- There is a strong opposition among NGOs in DCs against JI.[16] Through JI the imbalances between ICs and DCs such as the excessive consumption patterns in ICs, might not only remain intact, JI might even extend them further to DCs.

  This problem of imbalances can probably partly be met by an administrative/organizational approach. A minimum requirement for an IC party eligible for JI should be set. This party should be able to give proof that it owns technology or that it has already achieved CO_2 emission reductions in its own facilities.
- The DCs' official channels and mechanisms should preferably be used during the preparation of, the negotiations on and finalization of the JI project proposal.

  In the ASEAN countries these projects should be attached or complementary to existing or forthcoming projects in the form of: energy efficiency enhancing programmes; and reductions of CO_2 or CH_4 emissions (from natural gas operations in their various stages or from sewage, waste, wastewater, solid waste, landfill, *etc.*).
- If JI is to be regarded as a direct foreign investment paid for by unused CO_2 capacities, through emission reduction or absorption, how can the maintenance or enhancement of CO_2 sinks unambiguously be valuated? This might become urgent if JI through reforestation projects develop further.
- In the FCCC it is stipulated that GHG emission inventories at a country level should be submitted. Therefore, quality and accuracy of the data play an important role. In this respect it should be recommended to standardize the methods to measure and calculate CO_2 emission inventories.

As has been pointed out earlier, neither the terms nor the conditions for JI have been clearly and unambiguously defined. Many interpretations and viewpoints have circulated, which has increased rather than decreased the scepticism of parties in DCs. Hence the DCs attitudes cover the whole spectrum of attitudes towards JI: from cautious welcome to confusion and from accepting to rejecting.

If the wholehearted participation of the DCs is desired, many of these controversies and incongruities should be solved or clarified. However, there are no indications showing that the energy policies of the ASEAN countries are in conflict with the correct execution of JI as long as the projects are in line with the laws and policies

---

[16] Sari, 1994.

of the DCs. Further clarification should however be sought on the various issues concerning the execution of JI, which have been discussed above.

5 Benefits from JI

Notwithstanding the sceptical attitudes of the DCs, JI can result in benefits for both the DCs and the investing countries. A potential benefit of JI could be, for instance, that the financing of the development and construction of power-generating plants, which has a high priority in the ASEAN region, could partly be carried out through JI. Negotiations on these types of projects should be carried out on equal footing and be kept transparent in all details. The benefits from JI for the DC should furthermore:

- be tangible and measurable;
- mean a positive contribution to CO_2 emission reductions by the main project; and
- contribute to its national priority and global commitment.

These benefits can be:

- *Financial.* JI can bring in additional project funds (investment for environmental purposes). Further financial benefits depend on the negotiated price of the unused CO_2 capacity.
- *Technological.* JI opens venues for advanced energy efficient environmental technology transfers from IC to DC and from an IC to a DC or from one DC to the other. JI can also provide opportunities for technology application on a wider scale (manufacturing of energy efficient equipment).
- *Social.* JI creates job opportunities and quality of life improvements.
- *Political.* if JI is executed it could contribute to the North-South dialogue and to the cooperation embedded in the FCCC.

The objective of CO_2 emission reduction in the energy sector is complementary to the main objectives of the energy policies of the DCs under review, guiding energy development planning and energy utilization that are essential to the DCs towards the turn of the century and beyond. JI can be a component of these activities. It is therefore important to practice JI as learning by doing. One can in this respect think of the *Offshore Production Sharing Contracts* in Indonesia, which have been observed and studied by other DCs, or of the treeplanting projects set up in South

East Asia. Through such offset projects, which elements are similar to JI, significant experience with JI projects can already be gained by potential parties.

6 Conclusions

- The energy policies of the DCs - as illustrated by the policies of ASEAN countries - show that provisional preparatory steps to control fuel combustion related CO_2 emissions have been taken.
- CO_2 emission reductions and stabilization should be complementary to the main objectives of energy policies: to make energy available and to use energy in progressive quantities so as to provision the developing economies.
- There are opportunities for JI in the ASEAN region provided that some necessary conditions (as indicated in this contribution) be met.
- JI together with development assistance can be applied in the same host country (DC) for the same projects as long as the objectives are complementary and clear.
- JI in the ASEAN region should best be handled through the official channels of the host countries.
- JI should be in line with the DC current policies.
- JI should not give the impression of being a rich man's programme which can buy and engineer everything.
- JI programmes should be used to enhance energy efficiency programmes or to introduce advanced environmental technologies.

Annex 1. ASEAN Energy/Gas Demand Centres.

	Area (Km^2)	Population (mln.)	GDP ($ bln.)	Energy Production (Ktoe)	Energy Consumption (Ktoe)
Thailand	513,000	56.9	92	14,400	23,600
Philippines	300,000	63.4	45	3,400	9,700
Malaysia	330,000	18.3	45	51,400	14,400
Brunei Darussalam	6,000	0.3	4	17,500	440
Singapore	1,000	2.8	46	0	3,100
Indonesia	1,905,000	186.0	116	144,800	40,000
ASEAN	3,055,000	327.6	348	231,600	91,500

Source: AEEMTRC, 1993b: Brunei Darussalam (p. 31); Indonesia (p. 49); Malaysia (p. 67); Philippines (p. 83); Singapore (p. 98); Thailand (p. 112).

Annex 2. ASEAN Energy Policies.

Table 13.1A. ASEAN Energy Policies: Brunei Darussalam.

Policy Objectives		Policy Measures	
1	economic optimization and reduced dependency on oil and gas reserves	1	tapping other sources of energy (*e.g.* minihydro and solar energy potentials)
2	engaging in energy diversification programme	2	encouraging private sector participation in energy development in the form of Build-Operate-Own (BOO)
3	promotion of the energy conservation programme	3	considering only the most efficient type of power plants (*e.g.* combined cycle)
		4	tariff revision to discourage energy wastage and create awareness of the high cost of energy
		5	promoting energy efficiency in building designs and choice of materials

Table 13.1B. ASEAN Energy Policies: Indonesia.

	Policies Objectives		Policies Measures
1	assure the availability of energy for domestic use at affordable prices	1	intensification by increasing and expanding surveys and explorations of energy resources which include oil, gas, and others
2	make energy available for export	2	diversification or reduction of the share of petroleum in the overall energy consumption by developing and using non-oil resources
3	diversify energy sources for domestic use and reduce the role of oil	3	increase in oil price to reduce government spending on subsidy for oil
4	develop new and renewable energy sources	4	release of Presidential Decree (1991) enforcing the use of energy efficiently and wisely through public campaign and educational programme, regulation and labelling. The target is a reduction of 17% by the year 2000
5	protect the environment	5	certain measures of the energy conservation and diversification policy are directly related to the reduction of GHGs (CO_2)
6	strengthen national resilience		

Table 13.1C. ASEAN Energy Policies: Malaysia.

	Policy Objectives		Policy Measures
1	ensure adequate energy supply by reducing dependence on oil and by developing and utilizing alternative sources of energy	1	Four-fuel Diversification Strategy (1979) identifies coal, hydro, gas and oil as the primary energy sources to meet national energy requirements. This strategy includes the development and use of alternative sources of energy
2	promote and encourage the efficient use of energy and discharge of wasteful and non-productive patterns of energy consumption	2	National Depletion Policy (1980) safeguards the exploitation rate of the nation's oil reserves by restricting the production to a percentage of the oil initially in place
3	minimize environmental degradation in realizing the above goals	3	National Petroleum Policy ensures reasonable prices, eases development of petrochemical industries and ensures Malaysia's participation in the development of the sector
		4	deregulation of the power generation sector and encouragement of completion with the privatization of the electric utilities
		5	National Energy Efficiency Programme (1991) includes energy conservation awareness programmes, energy audits in buildings and industries as well as studies on fiscal incentives, transfer of technology, legislation, *etc*
		6	Environment Impact Assessment (EIA) study for all new energy projects
		7	encouragement to use unleaded petrol (introduced in 1990) and natural gas

Table 13.1D. ASEAN Energy Policies: Philippines.

	Policy Objectives		Policy Measures
1	provide adequate, timely, sustainable and reasonably-priced and equitably-distributed energy forms	1	sustain momentum in the exploration and development of oil and other indigenous sources
2	promote the judicious and efficient use of energy	2	diversify sources and types of energy imports
3	achieve energy objectives while maintaining ecological balance	3	enhance private sector participation in energy projects
		4	promote fuel substitution and diversification in power generation
		5	formulate and strictly implement power plant objectives and measures and rehabilitation programme
		6	provide electricity in commercially viable areas
		7	promote energy research and development and intensity promotion of energy conservation and energy-efficient technologies
		8	restructure tariffs to encourage efficient electricity use
		9	move towards downstream oil industry deregulation
		10	integrate environmental concerns in the planning and implementation of energy projects
		11	enhance assessment of and planning for the energy needs of countryside development

Table 13.1E. ASEAN Energy Policies: Singapore.

Policy Objectives	Policy Measures
1 pricing	**Pricing**
2 energy conservation	1 letting market forces of supply and demand determine energy price to ensure that energy is used more efficiently
3 efficiency in energy production	2 quarterly review of electricity and gas tariffs to ensure that they reflect current fuel prices
4 diversification in fuel supplies	**Energy Conservation**
	3 promotion of energy conservation through a combination of financial incentives and disincentives and through education
	4 incorporation of a set of energy conservation standards for building design into the Building Regulations
	5 design of fiscal policies to encourage the public to use smaller, more efficient vehicles
	6 encouragement of the use of public transport
	Efficiency in Energy Production
	7 achieving greater efficiency such as the generation of electricity through power station upgrading
	Diversification of Energy Production
	8 energy diversification through the use of natural gas in addition to oil

Table 13.1F. ASEAN Energy Policies: Thailand.

	Policy Objectives		Policy Measures
1	ensure availability of energy	1	entering into long-term energy contracts with neighbouring countries
2	promote free-market competition in the energy sector	2	deregulation of the petroleum product prices
3	implement energy conservation programmes	3	restructuring of electricity tariff in line with marginal cost
4	promote private investment in the country's energy affairs	4	liberalization policy of free market competition and privatization through sale of state enterprises in the financial market
5	reduce environmental impact from energy use	5	sale of electricity from the private sector to the national utilities
		6	launching of a Demand Side Management Programme to reduce both expected peak load and electricity demand
		7	promulgation in 1992 of the Energy Conservation Promotion Act which gives legal entity to the Energy Conservation Fund
		8	encouragement of production and use of more environment-friendly petroleum products (unleaded gasoline)
		9	regulation of environmental protection equipment for coal-fired power plants and later on for cars

References

AEEMTRC, 1993a. *ASEAN 2010 Program*, ASEAN synthesis report, Jakarta, Indonesia.

AEEMTRC, 1993b. *Energy in ASEAN Country Profiles*, Jakarta, Indonesia.

Blok, K., E. Worrell, R. Cuelenaere, and W. Turkenburg, 1993. 'The Cost-effectiveness of CO_2 Emission Reduction Achieved by Energy Conservation', *Energy Policy*, 21 (6), June 1993, pp. 656-61.

Hippel, D. von, P. Raskin, S. Subak and D. Stavsky, 1993. 'Estimating GHG-emissions from fossil fuel consumption 2 approaches compared', *Energy Policy*, June 1993, pp. 691-702.

Metz, B., 1994. *Joint Implementation as a Financing Instrument for Global Reductions in GHG-emissions*, in: A. Chayes and K. Ramakhrisna, *Criteria for Joint Implementation Under the Framework Convention on Climate Change*, Woodshole Research Center Workshop Report, January 1994.

Mintzer, I., 1994. *Institutional Options and Operational Challenges of a Joint Implementations Regime*, Stockholm Environmental Institute and Center for Global Change, in: A. Chayes and K. Ramakhrisna, *Criteria for Joint Implementation Under the Framework Convention on Climate Change*, Woodshole Research Center Workshop Report, January 1994.

Moomaw, W.R., 1994. 'Achieving Joint Benefits from Joint Implementation', in: A. Chayes and K. Ramakhrisna, *Criteria for Joint Implementation Under the Framework Convention on Climate Change*, Woodshole Research Center Workshop Report, January 1994.

Sari, A.P., 1994. *Solution or Problem? Joint Implementation in the Climate Convention*, February 1994. Prepared for the Indonesian Core Group of the Climate Action Network - South East Asia.

Sasmojo, S. et al., 1991. 'CO_2-emissions reduction by price deregulation and fossil fuel taxation', *Energy Policy*, December 1991, pp. 970-7.

Sasmojo, S. and M. Tasrif, 1993. *Energy Strategy Options of Developing Countries Coping with Global Climate Change Issue*, paper presented at Workshop on Global Warning Issues in Asia, organized by AIT, Bangkok, 8-10 September 1993.

Wisaksono, W., 1993. *Natural Gas: the Asia-Pacific Challenge; Conference on Regional Energy Strategies and Commercial Opportunities for Gas*, 8-11 February, 1993, Kuala Lumpur, Malaysia.

Wisaksono, W., 1994. *Natural Gas in ASEAN's Energy Mix at Present and in the Next Century*, 1994 Gas Industry Summit, Jakarta, 24-25 January 1994.

14. Joint Implementation

Cautions and Options for the South

R. SHAKESPEARE MAYA*

1 Introduction

Within the context of the United Nations FCCC effort the participation of developing country parties in emissions reduction policies is considered to be very important. Although they have no specific targets set for them, these parties' involvement in the global debate on climate change is desirable as they are about to enter the economic expansion process which led to the present level and rate of GHGs accumulation in the atmosphere when the present Annex I countries experienced the same.

For developing nations JI represents a serious dilemma. Initially, JI was viewed and appeared to take place among Annex I countries. Furthermore, developing nations viewed the JI proposition with a lot of scepticism and in some cases the option was totally rejected as 'environmental colonialism'. The position of developing nations has, however, since then changed. While still viewed with some scepticism JI is now considered an opportunity for technology transfer from the North to the South.

A characteristic of the logic of JI which has yet not been discussed is that the host country for the JI project should itself have excess capacity to emit. It is questionable whether this is a necessary requirement. Can two countries, each with exhausted capacity to emit but both required to reduce emissions, not enter into a JI arrangement and share the credits? The present discussions on JI appear to indicate that credits for JI accrue only to the *'guest investor'* and not to the *'host investor'*.

Whichever way the argument may go, JI appears to offer the greatest potential for international cooperation in reducing emissions of GHGs. It has been postulated that while the Global Environment Facility (GEF) wields only a couple of bln. dollars for supporting developing countries in carrying out emissions reduction activities, in the immediate future emissions reduction transactions under JI and corresponding financial transfers are likely to exceed some $150 bln. Even when viewed from an institutional perspective, JI is likely to be decentralized, allowing various parties to enter into independent arrangements and eventually reverting to some Central Authority for the accounting of credits. In contrast, GEF based activities have to go

* R.S. Maya is director of the *Southern Centre for Energy and Environment*, Zimbabwe. He is also lead author of IPCC WG III, writing team 7.

C. J. Jepma (ed.), The Feasibility of Joint Implementation, 209–217.

through a centralized bureaucracy which has the inherent effect of constricting activities and delaying actual emission reductions in the field.

2 North-South JI: Reasons for Scepticism

This section discusses the basis for scepticism and enthusiasm in developing countries with respect to North-South JI (NSJI). The primary reason for caution in developing countries is that they doubt the genuineness of the change of heart among industrialized nations particularly in view of the massive transfer of state of the art technology to the South that is possible under NSJI. This doubt emanates from the steadfast 'technology denial' stance taken by the North over the past decades. Within the framework of the North-South dialogue seeking to establish a New International Economic Order (NIEO), developed countries significantly failed to accede to the developing country demands for an equitable access to global technological development and an equitable sharing of world markets. Particular concern in the developing world was with unfair commodity terms of trade and general unequal exchange inherent in North-South economic cooperation.

This dialogue, which also took place under the auspices of the United Nations institutions, has failed to engender any meaningful transfer of technology to the South. Furthermore, in places where such transfers were attempted, results were pathetic for the simple reason that technological imports were not matched with the necessary access to markets in the North. This matching would have been necessary to enable the importer of Northern technology to derive foreign funds to pay for the technology.

The unequal exchange which prevailed since direct colonialism became termed neo-colonialism. The neo-colonial era saw the North achieving maximum gains in terms of resource transfer and labour enslavement in the South. As the South increased pressure against such unfair practices, the North made some concessions with respect to technology transfers but this was packaged in a new cooperation paradigm under the concept of Export Processing Zones (EPZ). The unemployment plight of the South created some vulnerability on the part of the South which was successfully exploited by the North. Under the EPZ formula, a host country simply provides land and basic amenities and exempts the investor from virtually all local regulations intended to protect workers against unfair labour practices, taxation and other local conditions essential for fostering local industrial development. Basically, the EPZ operations have no forward and backward linkages with local industrial development efforts except through the provision of cheap labour from local personnel.

While these forms of neo-colonialism have hardly been shaken off by the Third World, a new cooperation paradigm is being ushered in under JI. Assuming developing nations adopt a positive attitude and decide to give the new initiative 'a chance', what are the real practical issues to be clarified in order to ensure a meaningful exchange under NSJI? A number of specific prevailing conditions have to be highlighted. In the same light, a number of critical questions needs to be answered. Some of the prevailing conditions are discussed below and some of the critical questions are listed in box 1.

3 Current Conditions Which Concern Developing Nations with Respect to NSJI

NSJI has generated significant interest among entrepreneurs in industrialized countries. There are two main reasons for this interest and, to some extent, enthusiasm:

- JI inherently provides the polluting company in the North an opportunity to achieve reductions of emissions at lower cost. NSJI is even more attractive in this respect as costs are bound to be much lower in a NSJI investment than in a North-North JI investment. Industrialists in the North are banking on the historically unequal relationships where labour in the South was easier to exploit than in the North. So viewed from the Southern perspective, the exploitative mind-set of the Northern countries has not changed since they would only participate in NSJI if and only if those unfair conditions are maintained. The South can therefore not correctly view JI as a way out of the technological quagmire or as an instrument for fairer and more equal exchange achieving the overall goal pursued under NIEO.
- JI with the South provides Northern companies an opportunity to introduce both technology and finished products in new markets. Under a JI regime, there will be international environmental pressures - channelled through private investment, bilateral collaboration and through the activities of multilateral bodies - in favour of cleaner production and cleaner life-styles. These pressures will have the effect of persuading developing countries to more readily accept Northern products in their markets. Because these products are supposed to be cleaner than those currently available in both Southern and Northern markets, they are likely to be new and experimental or at least unable to survive in the market without support. The Solar PV dissemination project in Zimbabwe and the Compact fluorescent lamp project in Mexico (see also the contribution by Wilms and Luzuriaga in Chapter 24 and 25 respectively) are good examples of new cleaner but immature

(unable to survive on their own in the market) products being diffused under the banner of the FCCC.

In addition to the fact that developing countries have different views on these reasons for cooperation a further set of concerns exists for them.

The weakness befalling a local company in a host country in a NSJI would increase with the pressure applied on it through greater stringency from local environmental legislation and with the difficulty of access to international markets.

If a government in a potential host country imposes stringent environmental conditions on its own companies, the host company's lust for JI will also increase in order to meet its local environmental standards. Under this condition, the host company finds itself in a constrained and weak position to negotiate a JI project with a potential Northern partner. This is because the host company will be forced to take the initiative to invite a Northern partner, who now, knowing the interest of the host company will impose tougher conditions for a technology transfer.

Further, even if the host company did not face stringent local environmental conditions, it would still require cleaner production techniques in order to facilitate access to Northern markets where it now has to compete with more marketable products from cleaner production processes. Without the push from Northern partners, the local company would find it imperative to seek to produce at the same standards as those expected in the Northern market. So there is a mutual interest in NSJI between potential hosts and guests.

While the initiatives on NSJI are likely to be initiated by Northern private investor interests (based on reducing the cost of emissions reduction activities), participation in the South is likely to have greater involvement of official institutions. This is because development projects with significant impact on emissions reduction are still in the public sector. Highly experienced profit-motivated Northern private companies will be seeking to engage in joint projects with public sector institutions in the South. The latter, who are likely less experienced, will have a social motive and would therefore be keen to participate. This socially motivated development enthusiasm may have the effect of reducing stringency with incoming JI projects.

NSJI may also turn out to be involuntary on the part of Southern partners. This is because Northern governments wanting to support JI efforts of their investors may find it expedient to freeze or reduce all other forms of North-South cooperation thereby leaving JI as the only available form of such cooperation. In view of this possibility, developing country partners have argued for 'additionality' in the support for climate change interventions. It would therefore be expected that NSJI would not replace or diminish existing support for development under present multilateral or bilateral arrangements. Developing countries, however, are in a very weak position to verify the effect of NSJI resource transfers on present mechanisms since levels of bilateral support are basically discretionary.

4 JI and the Abatement Cost Curve

The GHG abatement cost curve (Figure 14.1) contains low cost options (on the left hand side) and high cost options as one moves up to the right hand side of the curve. It is quite possible that present results of abatement costing may be used as the initial indicator or basis for Northern partners to select activities for NSJI. This is likely to happen if Northern partners are allowed to invest in activities other than those in which they have specialized themselves in their Northern environment.

Box 14.1 Some critical questions on NSJI within the UNFCCC.

- What is the duration of validity for NSJI cooperation? How long will this cooperation paradigm last?
- What is the impact of NSJI on present arrangements to support Southern participation in emissions abatement?
- What is the extent of risks of engaging in NSJI?
- Who will pay for the transaction costs in NSJI projects?
- What will be the ownership arrangements for NSJI investments?
- How will credits be shared or distributed among Northern and Southern partners in NSJI?
- Will NSJI projects include replacements costs for out-dated machinery in developing nations?
- Will they represent new foreign direct investment or will they be a market creation gimmick for unproven Northern products?
- How can one determine that the technology being transferred is the best available or matches that in use in the Northern partner's factories?
- What will be the role of Government in developing private sector NSJI projects?
- How will it be ensured that Southern NSJI products will continue to enter markets in the North?
- Will the enthusiasm with NSJI create a debt burden as the North relaxes its conditions for technology transfer?
- What present developing country advantages will be wiped out at the end of the NSJI era?

For example, a detergents company in the North may see it fit to undertake 'joint implementation' through sink enhancement in the South - through a forest development project. Under this set up, the Northern partner would find it logical to invest in projects on the left hand side of the cost curve in a developing country as it would in its own environment. The effect of this 'selectivity' is to wipe out the cheaper cost options in developing countries through JI to the advantage of Northern partners and to leave only the more expensive right hand side options for developing nations to implement on their own in the future.

Figure 14.1. A typical abatement costs curve with negative cost options.

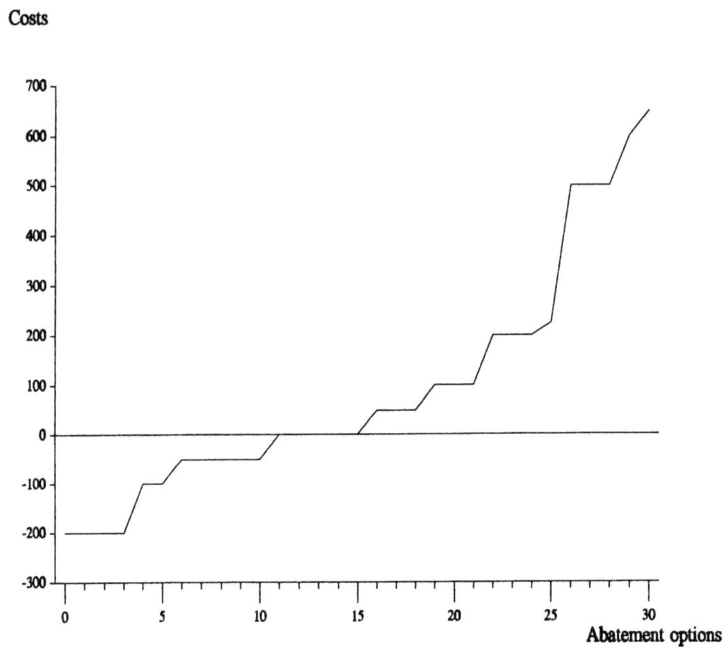

Such a trend would have two institutional effects. The first is a weakening of the presently valid arguments that developing nations have not made significant contributions to the present accumulation of emissions into the atmosphere. The second is an increase of the rate at which the weakening of the developing country position occurs.

In order to minimize the occurrence of such a trend, it may be useful to consider a situation in which a Northern partner only carries out JI projects in fields where it currently generates emissions. A manufacturing company, for example, would not be allowed to carry out projects in afforestation. This condition would first speed up industrialization in the South since Northern companies would have to transfer technologies in areas where they hold notable expertise.

It may also happen that Northern companies will only transfer currently known technologies. These technologies, however clean they may be, would become 'environmentally obsolete' as they would be overrun by newer and cleaner technologies resulting from increased international efforts to generate cleaner production techniques. If this happens, then the present generation of technologies transferred to the South will become unacceptable as would their products. Depending on how rapidly this environmental obsolescence occurs, developing nations will find themselves again groping for the North for the later generation technologies but this time on their own as the justification for NSJI would have been removed due to the fact that cost in either location would be basically pushed to parity as a result of the tendency to implement the lower cost options first.

If the obsolescence is gradual and long term, it should be expected that technology transfer under NSJI cooperation would have raised the competitive position in both camps to parity.

It is therefore ideal that *parity in costs* should coincide with *parity in competitiveness*. Engaging in NSJI without mechanisms to ensure this coincidence may lead to an undesirable result: Northern partners (at least the unscrupulous of them - and these are many if we judge from the 'toxic waste dumping' events of recent years) may find an opportunity in NSJI to dump technologies in the South and perhaps to engage in 'planned obsolescence' programmes whereby the introduction of cleaner technologies is programmed to create a perpetual series of short term markets.

5 Essential Developing Country Positions on NSJI

There is no coherent Southern Position on NSJI. Developing country positions differ by bloc of developing countries and perhaps by level of economic development. The latter cleavage is not so clear cut. The former can be defined through two observations. The first is that African bloc countries have no specific position on JI - they have not yet formulated a coherent or studied opinion. The Latin American developing countries seem to be prepared if not enthusiastic to enter into NSJI projects, but they do not seem to have systematically studied the form of engagements most advantageous to them. The Asian bloc countries also have accepted NSJI but still hold serious scepticism about this form of cooperation. These countries have tried to engage in the international debate on JI and to influence the definition of this new development paradigm.

Analyzed in the light of the African bloc countries, it can be concluded that whatever the direction JI is taking, Africa is basically being left out of the NSJI debate and NSJI investment. The reason for this may be a prevalence of the traditional investment scepticism with Africa based on arguments of poor infrastructure, instability and lack of human resources. Whatever the reason, the fact remains that initial NSJI engagements are targeting the other two developing country blocks - Asia and Latin America.

Ultimately, NSJI provides a good opportunity for technology and resources transfers to the South and to Africa. But this can only be a meaningful exercise for the continent if the criteria listed in box 2 are introduced and observed. Observing these conditions, African countries should actively seek to identify the benefits to them of NSJI and seek Northern collaboration to achieve development under this paradigm.

Box 14.2 Some Criteria for NSJI

- NSJI should be in an area in which the Northern private company wishing to cooperate is presently operating.
- Abatement options in areas where no Northern partners have present interest or experience should be left to bilateral or multilateral support - this should be the case with forestry or sink enhancement abatement option.
- Manufacturing sector NSJI should include marketing support packages.
- In cases of rapid environmental obsolescence of manufacturing technology, packages to ensure replacement access to new and cleaner technologies should be included in the NSJI arrangement.
- NSJI activities with a strong market creation orientation should not be accepted.
- NSJI interventions in economic areas where a strong public sector interest in developing countries exist should be left for bilateral or multilateral cooperation.
- NSJI should not be carried out through EPZs.

6 Should there be South-South JI (SSJI)?

SSJI has no basis. It is accepted that the South has made no significant contribution to the present atmospheric mayhem resulting in global warming and climate change. For this reason they have no reason to engage their own resources to reduce emissions - hence the argument that the North should pay the agreed incremental costs of Southern participation in reducing emissions.

SSJI may become necessary once Southern levels of industrialization have risen sufficiently to indicate that they are making significant contributions to GHG accumulations. As indicated earlier, the arrival of such a situation can be enhanced by cooperation under NSJI.

Even with the most optimistic arguments, it is difficult to paint a valid scenario where SSJI is meaningful. The closest condition for SSJI would be a situation where two abatement cost curves for two developing countries are compared and one is inherently more costly than the other. Under this condition, the country with the more costly curve may wish to reduce its abatement costs by cooperating with the lower cost country under a SSJI arrangement. But then, no developing country under the present arguments and understandings among the Parties would engage in abatement activities for their own sake outside the purvey of its own economic development. The country instead would seek to invest in economic development activities in the most 'emissions sensitive' approach possible. It would not transfer resources across boundaries simply to capture cheaper approaches for achieving this emissions sensitivity in investment.

15. Thoughts and Questions on Joint Implementation from the Perspective of a Country in Transition

TIBOR FARAGÒ and CSABA NEMES*

1 Introduction

The stabilization of the emission of GHGs is the most important commitment under the FCCC for the Annex I countries: developed countries and 'countries undergoing the process of transition to a market economy'. The introduction of the latter term in the text and the provisions of the Convention for the formerly centrally planned economies, which was a result of long debates, eventually marked the recognition of the special position of these countries.

Here only two features of the special circumstances with respect to the commitments of countries in transition will be mentioned. First, it is identified that these countries face, because of the transition they are in, different circumstances, *i.e.* extensive socio-economic changes, lack of financial resources, uncertainties about the structural changes of the economy, various environmental problems, *etc.* Second, the position of these countries in the context of their per capita anthropogenic GHG emissions differs from that of the other Annex I countries. With regard to the CO_2 emissions, the latter aspect means that although the per capita GHG emissions are in general considerably higher than in most of the developing countries or the worldwide average, these emissions are essentially lower than the per capita CO_2 release in most of the industrialized countries.

2 JI with Respect to the Countries in Transition

In the formerly centrally planned countries the inefficient economic structure resulted in a very high energy intensity: the energy consumption per unit of GDP surpasses three to four times that of the industrialized countries!

The successful solution of the tasks related to the transition depends *inter alia* on the extent to which these countries succeed in switching to a much less energy intensive economic structure. This switching process should consist of an adoption of advanced and environmentally sound technologies, in particular for those economic activities that are primarily responsible for the GHG emissions.

* The authors are associated with the Hungarian Ministry for Environment and Regional Policy.

C. J. Jepma (ed.), The Feasibility of Joint Implementation, 219–222.

These were the principal considerations in formulating our views on the FCCC. The considerations made us endorse the particular wording of the provision of the FCCC which emphasizes the importance of the facilitation of technology transfers (Article 4.5). In the declaration of the Hungarian government of 16 April 1992 this idea was already expressed:

> ...the effective implementation of this plan [*i.e.*, the CO_2-stabilization] and also the possible further reduction of these emissions considerably depends on the rate at which the country can stabilize, renew its economy and, in this context, how that energy efficiency develops and, furthermore, it depends on the economic and technological progress with the developed countries.

During the negotiations on the FCCC the Hungarian delegation admitted its due share in the commitments for the climate risk. As far as the overall effectiveness of the Convention is concerned, the Parties have, in principle, committed themselves to stop the Business-As-Usual (BAU) behaviour regarding man-made global emissions. However, as a matter of fact, they continue their intergovernmental discussions more or less as Diplomacy-As-Usual (DAU) resulting in a relatively slow process in responding to the climate risk. From the perspective of the countries in transition their current problems hamper their possibilities to strengthen the contribution to the solution of the climate change problem.

Because of the imbalance among the Parties to the FCCC, we should re-evaluate all possible approaches and tools to address the given environmental hazard. The mechanism of JI is just one of these potential instruments. In the countries in transition JI might be interpreted as a special framework for the cooperation with other countries in renewing their technologies and at the same time meeting the requirements of the FCCC. Besides this general interpretation, there are important questions which should be properly answered before JI can be applied.

If we now forget the provisions of the FCCC for a while, and think freely about the potential of the joint implementation of a wide range of relevant response policies, we might all agree that such an instrument may well serve to find the optimal or quickest ways to stabilization or reduction of the global GHG emissions.

It is evident that the optimal allocation of resources to realize the greatest reduction of net emissions can be achieved as cost-effectively as possible if we are not limited by geographical, sectorial, legal or other conditions. Therefore, if the mobilization of one unit of resources results in an significantly larger reduction in a different country or region, then - no doubt - it should be done purely from the perspective of global benefits. At the same time, there are various further aspects which should be carefully investigated to better understand the constraints and potentials of this mechanism.

The very low energy efficiency in all or most of the transition countries - one third or a quarter of the similar indicator for the industrialized countries as mentioned above - is one of the essential reasons which makes them, and in particular Hungary, very open minded to all forms of cooperation in the field of development, transfers or applications of advanced, environmentally sound technologies. This is because the 'transition phase' might involve the status of recipient as well as of donor.

3 Some Points of Concern

Turning back to the FCCC, we feel that the present framework does not provide enough encouragement to JI. Consequently, we should simultaneously act in two directions.

Firstly, in the spirit of and in accordance with the principal objective of the FCCC, we should return to the more holistic approach to address the anticipated climate change - which is a common concern. We can again think more freely about the various options to respond to this hazard.

Secondly, with the concrete provisions of the FCCC in mind, smaller scale concrete cooperative actions could be projected and jointly implemented to gather experience on the feasibility and efficiency of this instrument. It should be mentioned that we should carefully investigate the common criteria for JI as they will be decided upon by CoP 1. The current discussions on this issue have been started by the INC and it seems probable that such a procedure will not be formally used for the current commitments. More experiences are expected to be gained from a planned pilot phase which might be a very important experiment for the countries in transition as well.

Furthermore, the cost-effectiveness of such measures should be deeply analyzed. Provisional investigations indicated that countries in transition could offer specific areas for the realization of JI projects with relatively high emission reduction rates for the particular investments.

Another point of concern is related to the sustainability of the adopted technological measures which is at least partially coupled with the concept of intergenerational equity. This is a general problem with JI for all potential recipient countries. The mechanism of JI is a special alternative to the utilization of the tradable emission rights. Finally, we should also question whether the 'emission opportunities' for the next generations in our countries can at present be decided upon by us. It is already clear for us that development does not mean a steady growth in use of raw materials, energy, and a subsequent increase of waste and atmospheric emissions.

With these general ideas in mind, we are open to a further evaluation of this instrument and its possible application, at least during an extensive pilot phase.

16. Catalyzing a Market for Joint Implementation Projects

JIM LESLIE and SERGIO VERDUGO*

1 JI from a Business Perspective

1.1 Introduction

In this contribution, which will discuss JI from a business perspective, we hope to accomplish three objectives:

- first, we will describe the many actions that businesses and NGOs are currently taking to advance JI;
- second, we will show the need for all of us who see the opportunity of JI - who are stakeholders in JI - to work closely with our national governments to provide incentives for early action on JI; and
- finally, we will present our thesis that the effective international framework for JI will grow out of the decentralized action of many - rather than out of a centralized process of international negotiation.

1.2 JI and the Business Council for Sustainable Development

Before entering into the core of the discussion, we would like to preface our remarks by emphasizing that what energized us is that members of the Business Council for Sustainable Development (BCSD) see new opportunities for eco-efficiency, trade and technology cooperation through JI.

JI offers businesses an unprecedented opportunity for furthering global action to address GHGs. We now are at a unique moment in our global history, in which the imperatives of development needs in the South can be linked together with broad-based environmental concerns of the North. Therefore, within BCSD-corporations, businesses are moving to empower and engage every person in realizing their overall mission.

---

* Jim Leslie is Senior Vice-President of Corporate Services at TransAlta Utilities in Calgary, Canada; Sergio Verdugo is Planning Manager of the company CAP S.A. in Santiago, Chile. The authors are, as representatives of their companies, members of the Business Council for Sustainable Development, an international organization of senior executives dedicated to providing leadership on sustainable development, including JI.

C. J. Jepma (ed.), The Feasibility of Joint Implementation, 223–230.

JI furthermore offers an opportunity to complement the centralized processes of development banks and institutions as the decision-makers in development. With JI, the energy and efforts can be moved into direct business-to-business action, as we build on the creative capacities of many to meet the challenge of developmental and environmental performance. This process of decentralized action in relation to our global environmental and developmental concerns is now showing up.

Finally, through concepts such as JI, a unique opportunity arises to harness the workings of the marketplace in order to achieve environmental goals.

2 Actions in JI

2.1 The BCSD project 'catalyzing a market for JI projects'

Today, the technology to reduce GHGs and to sequester carbon is already available in the world, and a broad-based mobilization of action is taking place. More and more, business and NGOs are seeing that well-designed JI projects make good business sense, make good environmental protection sense, and make good sustainable development sense, so that many businesses are moving into action.

Whenever there exists an opportunity to add value through comparative advantage, people begin to trade with each other. They have never waited for a complete prescription of all of the detailed trading rules before acting on the opportunities.

Among the actions that are occurring today is a project which the BCSD is currently working on. This project, which is called 'Catalyzing a Market for Joint Implementation Projects', is under the direction of the authors of this contribution. As BCSD members, the authors are working together to stimulate action and to encourage national governments to move ahead rapidly with policies and incentives to make JI happen as part of their climate change action plans. The project is being implemented in three different ways:

- first, and most importantly, by working with the members of the BCSD and other business people as well as NGOs to support the development of good JI projects;
- second, by providing information to national governments and business communities on JI and the important new resources that JI can bring, especially to developing countries; and
- third, by developing a business position on how a framework for JI can provide incentives for business to take action.

Through the BCSD project, it has been discovered that business has already been extremely active in developing GHG offset projects and sorting out potential projects, all of which will provide valuable learning within the JI context.

2.2 The experience of TransAlta

TransAlta, by way of background, is a medium-sized, coal-based electric utility located in Western Canada. Due to its heavy dependence on low cost coal, it is one of the largest emitters of CO_2 in Canada and it therefore recognizes that management of GHG emissions is an important issue to address.

TransAlta Utilities has committed to stabilize its net GHG emissions at 1990 levels by the year 2000. This commitment has been made voluntarily by TransAlta as part of its commitment to sustainable development. TransAlta recognizes the need to take prudent, cost-effective action to address this issue. This goal will be achieved through a mix of options including: increased internal efficiency; promotion of customer efficiency; and the purchase of domestic and international offsets.

To ensure that projects are pursued in a consistent fashion, TransAlta is closely following a set of guiding principles established by a stakeholder advisory committee of senior representatives of environmental and international development organizations.[1]

2.3 There are many international JI projects

In Part IV of this volume there will be, among others, a review of the broad range of carbon offsets and initiatives similar to JI that are under development or already operational today. These include actions under the broad headings of waste management, energy efficiency, renewables, fuel switching, sustainable agriculture, methane reduction/recovery and forestry. We will just comment on a few random examples:

- New England Electric System has begun a carbon offset pilot project, working with a local forest products company in Malaysia to develop sustainable forest management practices;
- An international, US-based, not-for-profit development institution is beginning work with Indian, Bangladeshi, Tanzanian and Brazilian organizations to commercialize livestock feed supplements. The supplements are expected to increase

1 These principles were recently published in: TransAlta Corporation, 1993.

local farmers' incomes by improving the productivity of dairy animals, while reducing methane emissions from ruminant livestock;

- Resource Enterprises and Raven Ridge Resources Inc. in the United States are developing a methane recovery project with local partners in the Ukraine;
- A project in China is under review that has the capacity to increase overall production efficiency through integration of the power grid and retrofitting of thermal plants;
- Several projects are assessing the GHG impacts of sustainable agricultural practices in Central America and Canada, as well as the impacts of changing sustainable forest management practices in Russia, Latin America and Canada; and
- A consortium of interests is currently looking at the possibility of using renewable bagasse fuel for power generation in India.

While the list goes on (for a more extensive list, see for instance Trexler in Chapter 17, Part IV of this volume), the message is clear: there are many actions being taken or planned by businesses and other organizations. In so doing, they are creating the first actions of a new market that is delivering on the promise of reduced global GHGs, and is creating economic value and community development while dealing with an environmental concern.

3 The National Enabling Policy Framework

A multilateral treaty for JI may not occur for some time. In the mean time, however, JI remains an important international opportunity to create a cost-effective global system for managing GHGs. In order to support the creation of a future global JI system, it is important to develop national policies towards JI for which the experience and learning can be produced by action. The national governments, with input from business and other organizations, can begin to take leadership on this issue today.

Business and NGOs need to cooperate with national governments to develop national policy frameworks that ensure new opportunities for trade, technology cooperation and GHG reductions. This challenge will require a process that maximizes the contributions and innovative capacities of many participants.

A voluntary framework for climate change is one such process. Within a voluntary framework, a national government sets overall targets for stabilization and then challenges many participants in the economy, including corporations, to design their own customized and cost-effective strategies and actions to contribute to achieving these targets. These actions can be carried out domestically and internationally.

This framework should be designed so as to provide a stimulus for industry actions and to encourage a strong element of partnership between government and business. It should be developed based on the following principles:

- *inclusiveness.* The framework should include all GHG, all sources and all actions to enlarge the window of opportunities. All developed countries should include JI in their national action strategies for climate change. They should furthermore encourage JI under voluntary frameworks, such as the United States' Climate Change Programme (see also Graham in Chapter 12);
- *less rules, more innovation.* A framework that minimizes rule-making will make participation easier and stimulate more innovation. Under such a framework, industry should be allowed to design its own strategies and plans to reach overall targets; they should not be told how and what technologies to use to achieve those targets;
- *stability.* If industry can be confident that the rules governing a framework will be stable, it will be more interested in participating. Great care should be taken by the governments, as we go forward, to avoid penalizing industries that take voluntary actions in advance of a regulatory framework;
- *use of a registry.* Developed and developing countries should design registries that enable business to register JI projects. These registries should track and recognize projects that:
    - demonstrate verifiable net reductions of GHGs;
    - demonstrate clear economic and environmental benefits;
    - demonstrate support and agreement between the developed and developing country for a transfer reduction to the company in the developing country; and
    - show a verifiable agreement between the developed and developing country investors that all benefits, including the GHG reduction benefits, were accounted for when the deal was negotiated; and
- *a strong commitment by government.* Governments have an important role in facilitating innovative JI partnerships. The US Department of Energy is committed to supporting and funding joint international ventures that lead to demonstration JI projects. Other governments should consider similar ways of supporting business and others in pursuing JI opportunities.

4 A New Common Sense

Today, what we are seeing is the development of a new common sense. It draws its stimulus from the overall concern of global climate change and the targets that countries have set as a result of the FCCC. This common sense is reflected in actions by business and NGOs through JI to protect the environment, develop economies and support communities. Moving this common sense forward will challenge us all.

First, moving this common sense forward will challenge business to build coordination with local governments and continue action on real JI projects. To achieve the full potential of JI and to avoid the harsh reaction of stringent rules that limit cost-effective action, business should pursue JI projects that:

- begin with cost-shared demonstration projects to test viability within the evolving regulations of developed and developing countries;
- involve trading between companies that have a stated GHG reduction goal;
- are economically and environmentally sustainable and based on market principles;
- increase energy and process efficiency in developing countries through technology cooperation;
- encourage technology cooperation between developed and developing countries; and
- accelerate sustainable development in the South.

Second, moving this common sense forward will challenge governments to develop their own national agendas and enabling policy frameworks for JI. As part of the challenge, it is recommended that:

- acceptance of JI be included in the voluntary frameworks of national action plans in developed countries;
- in countries with fiscal or market-based economic instruments for GHG, JI projects be recognized as providing reductions in net GHG emissions; and
- developed and developing countries alike design national registries to record JI actions and their results.

In the absence of effective national incentives for JI, the burgeoning interest and enthusiasm for JI will be lost and will revert to the higher-cost action that will be a barrier to national and global sustainable development.

Finally, this new common sense will challenge the academic community to new research and knowledge inputs into the process of JI. Through new research activities and improving the knowledge on economic, legal and institutional aspects with respect to JI, the academic community can provide important inputs into the implementation of JI.

It can be concluded from this contribution that, at this stage of developing activities to curb climate change, it is important that the business community, national governments, intergovernmental bodies, and the academic community strongly cooperate in order to create an economy and a way of relating to our material world that is not an either/or argument, but a means to create the best life for the greatest number of people.

In other words, as put forward by Hawken:

> ... the globe needs to find a path that recognizes limits, while using our innovative capacity to invent and re-imagine our world to increase efficiency, decrease harm, and improve our existence (Hawken, 1993).

References

Hawken, P., 1993. *The Ecology of Commerce; a declaration of sustainablity,* First Edition, Harper Business, Harper Collins Publishers, New York, USA.
TransAlta Corporation, 1993. *Pursuing GHG offsets*, a report by TransAlta Corporation with input from the GHG Offsets Advisory Committee, Calgary, Canada.

PART IV

GREENHOUSE GAS ABATEMENT AND CARBON OFFSET STRATEGIES: COST STUDIES AND PILOT PROJECTS

17. Carbon Offset Strategies

A private sector perspective

MARK C. TREXLER*

1 Introduction

Globally, carbon offset funding has the potential to provide a new and large source of funding for energy and natural resource programmes around the world. Using funding from private companies in industrialized nations, for example, developing countries can gain access to more efficient and cleaner power generation and consumption technologies or can fund forestry and other natural resource projects. The addition of carbon offset funding to existing international funding sources can promote economic development, particularly during a time of constrained governmental resources for international assistance projects.[1]

The next two years are likely to be important for the development of carbon offset policy, particularly as it relates to the mechanism of Joint Implementation (JI). Although the ultimate form of future climate change mitigation policy remains conjectural at this stage, carbon offsets appear likely to constitute an element of any final greenhouse gas (GHG) mitigation policy implemented in the United States or other OECD countries. Carbon offsets, whether involving CO_2 or other GHGs, offer tremendous opportunities for increasing the economic efficiency of a GHG mitigation policy; additionally, carbon offset projects provide a vehicle for multiple groups to cooperate to achieve a range of separate or shared objectives, both environmental and economic. A well-designed carbon offset effort should be able to be characterized as a 'win-win' situation with all participants standing to gain from the effort.

2 A Background to Carbon Offsets

Discussion of the carbon offset concept dates back to at least 1977 when Freeman Dyson first quantified the area of new forest or tree plantation that would be required to absorb the quantity of fossil-fuel carbon being emitted through combustion each year. He concluded that 730 mln. ha. of fast growing trees, an area roughly equal to

* Mark C. Trexler is associated with Trexler and Associates, Inc. Portland, USA.

[1] Trexler and Meganck, 1993.

C. J. Jepma (ed.), The Feasibility of Joint Implementation, 233–248.

80 % of the land area of the entire United States, would be required to offset the emission of some 5 Gtons of carbon per year.[2]

Today, discussion of carbon offsets is much broader based. Development of viable carbon offset projects can be based on the off-site pursuit by a CO_2 emitter of any of three broad approaches:

- *Avoiding production of CO_2 in the first place.* These options include improving energy production and conversion efficiencies in another geographic area, preventing the loss of threatened forests, changing land-use management practices to conserve biomass, or subsidizing the switch from carbon-intensive fuels such as coal to less carbon-intensive fuels such as natural gas or renewable energy sources. With the tremendous expansion in the energy production and distribution sectors that is expected to occur during the next two decades and the tens of mln. of ha of forest that are expected to be lost, the aggressive pursuit of these options can significantly reduce long-term global CO_2 emissions. Although public policy reform also has a major role to play in promoting emissions reductions in these areas, specific 'carbon offset' interventions at the project level can be developed.
- *Removing CO_2 from the atmosphere after it has been released.* These options include planting new trees, improving management and growth rates of existing trees, or changing agricultural practices to increase soil carbon uptake. It is well-known that substantial amounts of land around the world that once supported tree cover could do so again; theoretical land availability is estimated in the hundreds of mln. of ha in both temperate and tropical zones. A great deal of this land is highly erodible or otherwise environmentally fragile; a restoration of tree cover is often its best long-term ecological and economic use.
- *Reducing the emissions or production of other GHGs.* Options such as coal-bed methane capture, landfill methane capture, reducing natural gas pipeline leakage, reducing biomass burning, or reducing agricultural fertilizer application and resulting N_2O production can be pursued. Although these options have not been as well studied as others, there is clearly a great deal of carbon offset potential here as well.

In recent years, more and more attention has focused on this range of options. More and more carbon offset projects are underway or are being considered. All of the offset options discussed above can have significant ancillary benefits beyond climate change mitigation, thus extending their 'win-win' potential. Energy efficiency

---

[2] Dyson, 1977.

measures can conserve natural resources, save money, and benefit national trade balances. Forest protection management and reforestation can promote biodiversity conservation, watershed protection, soil conservation, and other environmental and economic objectives.

Although there is little doubt regarding the conceptual validity of these three approaches to reducing long-term GHG emissions, there is still no accepted definition of valid carbon offsets, nor is there an accepted method of quantifying an offset's magnitude. The United States and some other countries have extensive experience with controlling other air pollutants, including through the use of offsets.[3] Considerable effort has been devoted to the development of offset programmes for such pollutants as NO_x and SO_2. These programmes are intended to generally ensure that offsets are:

- real, *i.e.* that pollution is actually decreased;
- permanent, *i.e.* that the reduction remains valid over a required lifetime;
- incremental or surplus, *i.e.* not required for compliance with other environmental regulations and not anticipated to regulations that would otherwise occur during the lifetime of the measure; and
- verifiable and enforceable.

CO_2, however, is quite different from most of the pollutants with which we have regulatory experience. Unlike NO_x and SO_2, for example, CO_2 is not an immediate health or environmental threat to human populations or sensitive ecosystems. Additionally, CO_2 does not result from impure fuels nor inefficient combustion; rather, it is the unintended byproduct of the energy production process. Also unlike NO_x and SO_2, CO_2 cannot be practically removed from flue gas prior to being emitted to the atmosphere.[4] Particularly important for offset purposes is the fact that CO_2 is a global gas and that it can be removed from the atmosphere through human intervention (using photosynthesis) once it is emitted. All of these factors should be taken into account and taken advantage of as rules and regulations governing carbon offsets and JI are developed.

---

[3] Marchant, 1992.

[4] Albanese and Steinberg, 1980; Fluor-Daniel, 1991.

3 A Review of Ongoing Offset Projects

3.1 Introduction

The first practical application of carbon offsets was the decision by a US independent power producer (IPP), AES Corp., to voluntarily offset the emissions of a future coal-fired power plant in the state of Connecticut through a tropical forest conservation project in the country of Guatemala. Today, as mentioned by Leslie and Verdugo in Chapter 16, many more utilities and other CO_2 emitters are investigating or are actually pursuing CO_2 offset projects involving several of the potential approaches discussed above. Some companies are experimenting with the concept to increase their understanding of, and therefore their ability to implement, available options at a larger scale in the future; others are responding directly to political pressure to be seen as doing something. Still others are responding to provisions beginning to appear in power purchase agreements.

United States utilities and IPPs actively pursuing carbon offset projects include PacifiCorp, New England Electric Services, AES Corp., Tenaska Inc., and Wisconsin Electric Power Co. Many other companies are currently investigating offset options. Internationally, the FACE (Forests Absorbing Carbon Dioxide Emissions) Foundation in The Netherlands has instituted the world's largest carbon offset programme, with projects underway in half a dozen countries in Europe and the tropics. In Canada, TransAlta Utilities is also undertaking carbon offset activities (see Leslie and Verdugo, Chapter 16).

Carbon offset activities can be differentiated based on their sources of financing. One category of projects comprises those sponsored by the private sector. A second category of projects comprises those being pursued by governments or intergovernmental organizations. The involvement of the government of Norway in an energy efficiency project in Mexico and a fuel-switching project in Poland may be a prelude to many more government-to-government carbon offset projects (see also Roland in Chapter 26). Similarly, the Global Environment Facility (GEF) is funding a significant number of projects as part of its global warming mitigation mandate.

It is the private sector projects that are the primary focus of this paper. This is because the interests and perspectives of governmental and intergovernmental funders of carbon offset and JI projects are likely to prove quite different from those of the private sector, and merit separate consideration. The issue of funding incrementality, for example, is a much higher profile one in the case of government funded projects than in the case of private sector projects.

Within the category of private sector JI, projects that are currently being pursued or which are in an advanced state of planning can be grouped into categories based on their technical approach. Not all of the projects briefly introduced below are international in nature. Whether a project is domestic or international is of little relevance to understanding the perspectives and needs of private sector funders in a future JI system.

3.2 Energy supply and demand efficiency improvements and fuel switching

Bynov Heating Plant - Czech Republic

Three US utilities have announced their intent to assist the Bynov Heating Plant in Decin in the Czech Republic to reduce GHG emissions by converting the heavily polluting, inefficient, brown coal plant to natural gas and making much needed energy efficiency improvements to the community's central heating system. This project which has elaborately been discussed by Helme and Gille in Chapter 10 is also a fuel switching project.

3.3 Forest protection and management

AES Corporation - Guatemala

As previously noted, the first carbon offset project was pursued by AES Corp. in Guatemala based on recommendations from the World Resources Institute, an environmental policy research organization in Washington, DC. The forestry components of the effort consist of tree planting in woodlots and agroforestry applications, increasing biomass yields through soil conservation techniques, and conservation of forest biomass through non-planting activities such as fire prevention. Farmers on small holdings are being trained and supported in planting trees to halt erosion, increase agricultural productivity, and increase wood supplies.

The World Resources Institute estimated that the project would more than offset the 15 mln. tons of carbon to be emitted over the life of a new AES power plant in Connecticut.[5] AES is spending $2 mln. out of an estimated project cost of $15 mln. As the first carbon offset project in the field, it has proven extremely valuable as a learning tool for the design of future offset projects. In particular, the project illustrated how difficult offset quantification and verification can be for some types of projects, and how stark the tradeoff between ancillary benefits and 'offset quality' can be.

---

[5] Trexler, Faeth, and Kramer, 1989.

AES Corporation - Paraguay

To offset a second power plant, AES Corp. is funding establishment of a nature reserve in Paraguay. A forest parcel targeted for sale to timber interests and containing endangered species has been purchased and is being managed for long-term conservation by The Nature Conservancy. The project was designed to offset the CO_2 emissions of the AES Barbers Point 180 MW coal-fired facility on Oahu, Hawaii. The International Finance Corporation (IFC) agreed to sell the 225-square mile parcel significantly below its market value of $5 - $7 mln. to assure its preservation. Known as the 'Mbaracayu' tract, the property is one of the last large tracts of dense humid sub-tropical forests in Latin America. Funding for the $2 mln. purchase was provided by the US Agency for International Development and AES Corp.

Tenaska Inc. - Costa Rica

Tenaska Inc., another US IPP, is developing a forest preservation project originally proposed by the government of Costa Rica. The project involves the purchase of private holdings within a Costa Rican national park. The holdings pose a major threat to the long-term survival of the park because the government cannot regulate land-use activity on the holdings under a recent Costa Rican Supreme Court ruling. Tenaska is intending to commit approximately half a mln. dollars to the project, with an expected benefit of 100,000 - 150,000 tons of carbon, based on a limited amount of matching funding associated with the goal of biodiversity conservation. Several mln. dollars would be required to purchase all the private holdings.

AES Corporation - Peru, Ecuador, and Bolivia

To offset the emissions of a third power plant, AES Corp. will support indigenous groups from Peru, Ecuador, and Bolivia in their efforts to gain control over their lands and develop sustainable resource extraction plans for the forest. In principle, protection of these groups' rights through legal channels will help prevent encroachment by outsiders and the subsequent conversion of traditional lands from tropical rainforest to grazing or agricultural uses. The challenge of monitoring and quantifying such an offset, however, is very great. AES funding for this combination of project efforts is budgeted at approximately $2 mln.

New England Power Co. - Malaysia

In Malaysia, the New England Power Company and Innoprise, a leading Malaysian forests products corporation, are involved in a project designed to offset carbon by reducing the carbon released during the logging process. In this pilot programme, Innoprise will implement improved harvesting techniques on 1,400 ha (3,500 acres).

Elements of the project include improved siting of logging trails, directional felling of trees, and vine removal prior to harvest. These practices are intended to decrease damage to undergrowth and unharvested trees during the logging process, thereby decreasing carbon released to the atmosphere and facilitating forest regeneration. New England Power Co. will spend $450,000 on the three-year effort. Pilot project reductions are estimated to be 100,000 - 150,000 tons of carbon.

3.4 Reforestation

A number of reforestation-based carbon offset projects are currently underway in the United States and elsewhere. These projects range considerably with respect to scale, species, and rotations.

PacifiCorp - Pacific Northwest

PacifiCorp is pursuing several pilot-scale carbon offset projects in southwestern Oregon and southeastern Washington. The projects involve pasture reforestation as well as reforestation of recently burned lands where natural regeneration is not expected to occur. Species involved include Douglas fir and Ponderosa pine. PacifiCorp's pilot projects cover almost 500 ha of private, non-industrial land. The company has developed a cost-share programme, modeled on federal reforestation programmes, that is premised on an agreement from the landowner to maintain the trees for between 65 and 80 years depending on the specific site involved.

Tenaska, Inc. - Pacific Northwest

Tenaska, Inc. is planning reforestation of almost 2,000 ha of land in southeastern Washington state as a carbon offset to complement its Costa Rica project. The mechanisms being used are very similar to those being used by PacifiCorp as described above, and are expected to yield carbon offsets at approximately $2/ton.

FACE Foundation - multiple locations

The FACE Foundation in The Netherlands, established by the Dutch Electricity Generating Board in 1990, has undertaken a series of reforestation projects in Europe and in the tropics. FACE plans to reforest at least 1,000 ha per year through up to eight reforestation projects in different countries, offsetting over the projects' lives the equivalent emissions of a 600 MW coal-fired power plant. The budget for this effort is reported to be over $10 mln.[6] One of FACE's projects is in Malaysia;

---

[6] FACE, 1994.

FACE has negotiated a contract with Innoprise Corporation, a state-run forest products enterprise, to reforest 3,200 ha of land in Danum Valley, Sabah, Malaysia before July 1995. If the project is successful, FACE will rehabilitate 25,000 ha in Malaysia over the next 25 years. Other FACE projects are located in Indonesia, Uganda, Ecuador, Poland, The Netherlands, and the Czech Republic. These projects will be discussed in more detail by Verweij in Chapter 23.

Russian reforestation
Several utilities, in cooperation with the US Environmental Protection Agency, are investigating potential carbon offset reforestation projects in several parts of Western Russia. The project would include the formation of a multi-party, intergovernmental partnership among agencies of the US government, the Russian federal and district governments, and various nongovernmental organizations.

3.5 Other greenhouse gas projects

Dairy cow methane reductions - India
Strategic supplementation of the diets of milk-producing animals can promote more efficient digestion and thereby decrease methane generation per unit head of cattle. One such project is currently being developed for carbon offset purposes in India. The project involves research on molasses-urea block formulation and process technology issues, production of high quality blocks, and an intensive marketing effort that involves recruiting and training village-based agents (predominantly women) to promote and sell the blocks and eventually operate as self-sustaining small-scale entrepreneurs. Details of the project have not yet been released by participating utilities.

3.6 Offset options yet to be explored

Several types of offset projects have not yet been pursued, most obviously the deployment of biomass energy and other renewable energy technologies. This is likely due to the relatively high per ton costs of pursuing such projects for carbon offset purposes.

4 Learning from Existing Offset Experience

As illustrated above, there are quite a few carbon offset projects currently being pursued for climate change mitigation purposes. It is notable that many of the projects profiled above are in the forestry sector. This can be explained in several ways. First, utilities and IPPs have been attracted by the ancillary benefits associated with forestry sector projects. Second, it is possible to pursue forestry projects at scales that are considerably smaller than those involved in most energy-sector projects. One can reforest 100 or 1,000 ha; it is much less easy to repower a small part of an existing power plant, or to fuel-switch just a small part of a plant's feedstock. Third, forestry projects appear to be particularly cost-effective from a carbon offset perspective.[7] These three explanatory variables, and the associated private sector motivations they reflect, should be borne in mind as debate over a JI mechanism continues.

There are many more projects, not profiled above, that proponents are likely to attempt to characterize as falling within the JI fold in the future, even though their original objectives had nothing to do with JI or climate change mitigation objectives. Indeed, developing a methodology by which to differentiate JI projects from others is important if the concept of JI and indeed of carbon offsets generally is to retain any meaning. The alternative could easily be to see almost every instance of foreign investment being termed an example of JI. This serves the long-term interests of neither the private sector, for which maintaining the political credibility of JI is a primary objective, nor of developing countries, for which a clear line of separation between existing aid and investment flows and JI is a primary concern.[8]

Even if one focuses just on projects where the climate change mitigation objective is a dominant variable, as is the case in the projects described above, it is far from easy to compare ongoing projects. Because there is no consistent definition of a good 'carbon offset' or of the criteria by which projects should be evaluated, the projects being pursued range dramatically with respect to their adherence to evaluative variables such as credibility, reliability, measurability, and verifiability. Nevertheless, the experiences of the companies undertaking projects thus far suggests several preliminary conclusions for private sector participants:

---

[7] Trexler and McFall, 1993.

[8] Trexler, 1993.

- There is no shortage of technical interventions in the energy or land-use sectors that can legitimately be characterized as carbon offsets due to their potential to decrease GHG emissions from otherwise prevailing levels.
- Conclusively demonstrating the magnitude of a carbon offset and establishing a no-project emissions baseline against which to measure a project's impact poses daunting technical difficulties that ultimately will require a policy-based solution. In the long-run it makes little sense to have to develop project-specific emissions baselines.
- Government involvement in private sector offset projects can both advance and hinder project objectives and viability.
- There is a general ignorance in many sectors of the needs and perspectives of private sector companies pursuing carbon offsets as a pollution control measure. In particular, there is little cognizance of the differences between carbon offset funding, in which project funders will likely be held accountable for the long-term success of a project, and conventional technical assistance or foreign aid funding where such accountability is far less prevalent.
- The political and academic debate over CO_2 mitigation options has confused many people regarding the context within which to think about carbon offset economics. It is easy to find literature suggesting that the cost of CO_2 controls can exceed $100/ton of carbon; it is similarly easy for individuals unfamiliar with the debate to conclude that these figures establish some sort of cost range for carbon offset projects. This is of course not true. First, in a voluntary emissions control system participants will never be able to pay such sums for carbon offsets. Second, there are still many inexpensive offset options available, both domestically and internationally.

The future costs of carbon offsets, inside or outside of a JI mechanism, are of paramount concern to all parties to this issue. Although many carbon offset opportunities are available, projecting the future costs of offsets remains extremely difficult. The costs of currently ongoing carbon offset projects vary widely, partially because the projects differ so markedly with respect to the credibility and reliability of the carbon offset itself.[9] The number of offset projects currently being pursued is simply too small to allow generalized conclusions about the real shape of carbon offset supply curves, either domestically or internationally. Generally speaking, however, carbon offsets should prove less costly than on-site emissions reductions for utilities and other CO_2 emitters, especially as the less expensive on-site opportunities

---

9 Kinsman and Trexler, 1993.

Figure 17.1. Identifying good carbon offset opportunities.

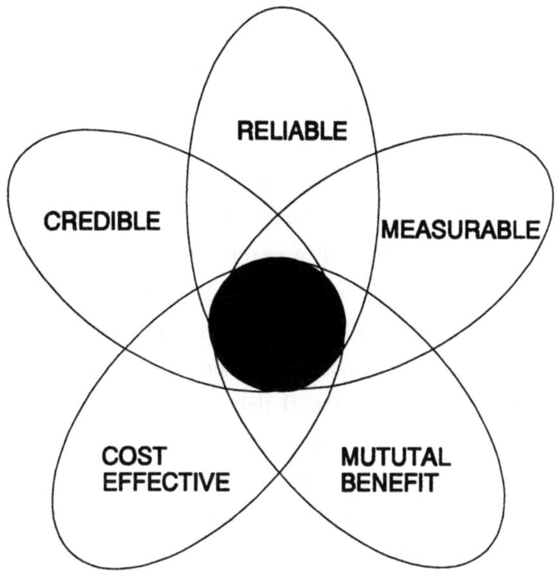

are taken advantage of.

5 From Carbon Offsets to JI

Notwithstanding the difficulties outlined above, well-designed carbon offset projects can be relatively simple. As shown in Figure 17.1, it is not difficult to work with a limited number of key objectives to identify projects that can advance the interests of funders and project hosts while maintaining project credibility, reliability, measurability, and cost-effectiveness.

The prospect of integrating relatively straightforward carbon offset opportunities into an internationally negotiated JI framework may or may not proceed as smoothly as past projects have. There are many things for private sector observers to the ongoing JI debate to be concerned about:

- *politicization of the JI debate.* This politicization could easily constrain the carbon offset options that can be pursued, significantly increase the cost of carbon offsets, and delay the development of viable JI guidelines.
- *establishment of JI clearinghouses.* Given the financial flows potentially at stake, several international organizations are beginning to compete for the opportunity to

manage these flows and to serve as a clearinghouse for future JI projects. Notwithstanding apparent economies of scale, many questions can be raised about whether the goals of effective CO_2 emissions reductions can best be met through the creation of large new carbon offset clearinghouse bureaucracies at the international level. Politics often play a major role in the allocation of funding through these international organizations and implementation follow-through has been a consistent problem since their inception. It can be difficult for private sector funders, who are concerned about being held accountable for whatever projects they get involved in, to work with governmental and intergovernmental bureaucracies which are very hard to hold accountable. At the same time, clearinghouses serving a more limited role in information exchange could be very valuable.

- *misunderstanding regarding motivations.* There is a fundamental misunderstanding among many participants to the JI debate regarding the options available to be pursued through the JI mechanism and the motivations of project funders. This misunderstanding is reflected in concerns that JI will result in a new environmental colonialism in which developing countries will be turned into carbon repositories with few opportunities for economic development.
- *absence of key groups of actors in the JI debate.* Neither the forestry nor the development communities have been significantly represented in the JI political debate to date. As a result, the benefits associated with important categories of potential JI projects are being undervalued.
- *lack of consensus.* There are many views regarding what JI is and what it should accomplish. Until there is some consensus about what JI is intended to accomplish, it will be impossible to agree on rules on how to achieve these aims.
- *uncertainties regarding standards.* There is the possibility that governmental and non-governmental suspicions of JI will lead to overly stringent standards for quantifying and monitoring the carbon offset benefits of JI projects. One can easily design monitoring and verification protocols that are so intensive and complex that they can exceed the basic cost of the project itself. An overly aggressive regulatory approach could also prevent pursuit of many economically and environmentally beneficial projects that might be slightly more risky than others.
- *unrealistic expectations concerning the private sector.* There has been an absence of recognition of what can and cannot realistically be expected of private sector funders as part of a voluntary emissions control regime. The political debate thus far has focused almost exclusively on issues pertaining to protecting developing countries from various kinds of economic and environmental exploitation. There has been very little discussion of the need to make JI friendly to funders as well, particularly when the participation of private funders in JI is largely voluntary.

Figure 17.2. Conflicting objectives could make 'good' JI projects difficult to find.

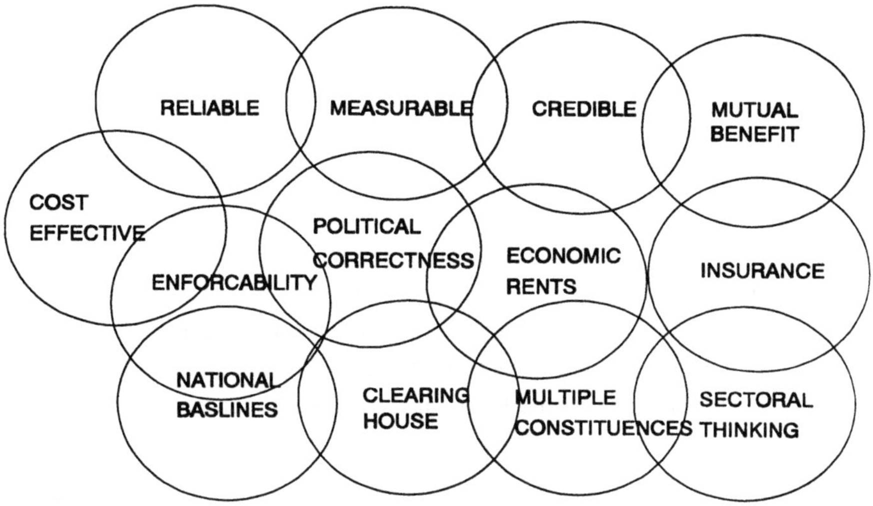

The net result of currently ongoing JI negotiations could look more like the confusion reflected in Figure 17.2 than the straightforward confluence of objectives found in Figure 17.1. If JI projects have to go through multiple screening processes, satisfy several constituencies, generate considerable economic rent for the host country, and have their success tied to national emissions mitigation performance against a baseline, JI projects capable of satisfying all of these restrictions as well as meeting the objectives of the funding entity may become hard to find.

As carbon offsets move into the intergovernmental arena of JI, fundamental policy truths need to be considered and addressed. Good carbon offset projects should be 'win-win' propositions, but this is not equivalent to asserting that all projects benefiting developing countries would make good carbon offset projects. Much of the need for economic development in the tropics, for example, occurs in areas characterized by aridity, climatological instability, high population densities resulting in unprecedented land pressures, uncertain land tenure, and unstable social, political and economic systems. Development projects in these areas are often process- rather than outcome-oriented. The process of encouraging community involvement and alleviation of suffering can be as important to project success in these areas as quantifiable deliverables. There may in fact be little emphasis on short-term deliverables because the social attributes being promoted - education, health, income - are long-term objectives.

It would be difficult if not impossible to structure good carbon offset projects around many of the economic development projects people would like to pursue in many of the circumstances just described. Potential carbon offset regulatory criteria requiring verifiability and measurability of a project's carbon offset benefits make them inappropriate in situations of great uncertainty or unpredictability and will conflict with the design of many economic development projects. Indeed, some organizations known for their economic development work are questioning whether to become involved with carbon offset projects precisely because they fear losing the flexibility to pursue their social and economic development objectives.

The fact that carbon offset projects are not a panacea for economic development needs to be recognized from the beginning. As public and corporate policy develops in the area of global warming mitigation, it may be that different kinds of climate change mitigation projects should be pursued by different funders. Private funders, for example, might best pursue projects specifically meeting needed regulatory criteria and providing for relatively secure long-term carbon benefits (*e.g.*, renewable energy projects, fuel switching, and reforestation). Public funders might best pursue projects that have high potential payoffs for both global warming mitigation and economic development but which may be difficult to reliably predict and verify in terms of long-term carbon benefit (*e.g.*, public education, agroforestry, and social infrastructure).

6 Conclusions

Large energy-producing or energy-consuming corporations in the industrialized countries are actively interested in carbon offsets as a climate change mitigation measure because of both the evolution of scientific thought regarding climate change and the growing consideration of mitigation objectives by legislative bodies. Most policy makers perceive energy-related companies, and utilities in particular, as primary targets for GHG emission control strategies. Thus, large CO_2 emitting industries such as electric utilities have a significant stake in finding viable cost-saving ways to meet future emissions restrictions or avoid emissions-based fees. Carbon offsets can offer emitters an alternative that is potentially cheaper than paying a carbon tax or directly reducing their emissions, particularly during this period of voluntary emissions controls when significant investments in climate change mitigation would be sceptically scrutinized by shareholders and regulators.

On the receiving side, climate change mitigation efforts are portrayed as a potential source of massive new funding for the energy and resources sectors of developing countries. The source of funding could be governmental (bilateral or multilateral) or non-governmental (private for-profit or not-for-profit firms). In today's budget-constrained world, however, governmental funding is more likely to result in a reallocation of spending rather than a significant net funding increase. Funding previously intended for economic development or sustainable forest management might be reassigned, for example, to a climate change mitigation budget line, even if the ultimate project being funded remains similar.

It is likely that private sector funding presents the greatest short- to medium-term opportunity for large-scale increases in funding for energy and resource management programmes intended to mitigate future climate change. Private sector participation in a JI process will likely remain largely voluntary for several years into the future, making it imperative that the needs of the private sector be accounted for during development of the JI process. These needs include the financial constraints associated with operating within a voluntary emissions control system and the need for corporate accountability in defending investments in carbon offset and JI projects. Yet it appears that the private sector could easily become caught between pressure from industrialized nations, which are looking to voluntary corporate actions to help achieve national emissions reduction targets, and developing countries, which are looking to private sector JI funding to make up for previous funding disappointments associated with bilateral and multilateral aid and to private sector funders to compensate for the GHG emissions sins of the industrialized North. It is not clear that private sector funding can live up to these expectations or that the private sector will choose to play in the JI game at all unless the rules are conducive to its participation. This is the challenge for JI negotiators and other interested parties during the next several years.

References

Albanese, S.S., and M. Steinberg, 1980. *Environmental Control Technology for Atmospheric Carbon Dioxide: Final Report*, Brookhaven National Laboratory.

Dyson, F.J., 1977. 'Can We Control the Carbon Dioxide in the Atmosphere?', *Energy* 2:287-91.

Fluor-Daniel, 1991. *Energy and Economic Evaluation of CO_2 Removal from Fossil-Fuel Fired Power Plants*, IE-7365, Electric Power Research Institute, Palo Alto, California.

FACE Foundation, 1994. *Annual Report 1993*, 20-3.

Goldberg, D.M., 1993. 'United States Should Meet its Year-2000 Goal Without Joint Implementation', *Energy, Economics, and Climate Change* 3(4):8.

Gowen, M., M.C. Trexler, M. Mendis, and L.H. Kosloff, 1994. 'Joint Implementation: Perspectives and Opportunities From the South and North', presented at: *Global Climate Change: Science, Policy, and Mitigation Strategies*, Air & Waste Management Association, Phoenix.

Kinsman, J.D. and M.C. Trexler, 1993. 'Terrestrial Carbon Management and Electric Utilities', *Water, Air, & Soil Pollution* 70:545.

Marchant, G.E., 1992. 'Freezing Carbon Dioxide Emissions: An Offset Policy for Slowing Global Warming', *Environmental Law* 22:623.

Trexler, M.C., 1993. *Intergovernmental Negotiating Committee VIII, Notes Re: Joint Implementation*, unpublished internal report, Trexler and Associates, Inc., Oak Grove, Oregon.

Trexler, M.C., P.E. Faeth, and J.M. Kramer, 1989. *Forestry as a Global Warming Mitigation Strategy: An Analysis of the Guatemala Carbon Sequestration Forestry Project*, World Resources Institute, Washington, D.C.

Trexler, M.C., and L.H. Kosloff, 1993. 'United States should encourage foreign offsets to meet year-2000 emissions goal', *Energy, Economics, and Climate Change,* 3, 9.

Trexler, M.C. and K.S. McFall, 1993. 'Building a CO_2 mitigation portfolio: the future is now for carbon sinks', *Electricity Journal,* 6, 60.

Trexler, M.C. and R. Meganck, 1993. 'Biotic carbon offset programs: sponsors of or impediment to economic development?', *Climate Research,* 3, 129.

18. The UNEP Greenhouse Gas Abatement Costing Study

Implications for Joint Implementation

JOEL SWISHER and ARTURO VILLAVICENCIO*

1 Introduction

The rate of carbon emissions from the energy sector in most developing countries is presently low, in terms of both per capita and total emissions. The largest source of greenhouse gas (GHG) emissions from some of these countries is from land-use changes in the form of deforestation. In the future, however, energy-sector emissions could increase dramatically in the course of economic growth and development. To what degree should these emissions be limited, and who should pay for the reduction measures?

One source of information with which to address this question is the UNEP Greenhouse Gas Abatement Costing Study, an international project on developing a common methodological approach for assessing the costs of limiting GHG emissions. The project was initiated in 1991 in an effort to help establish a basis for national studies of the economic issues which would be accepted as objective, accurate and comparable between countries. The purpose of this paper is to interpret some of the recent results of this project with regard to the issue of Joint Implementation (JI). To explore the potential benefits and risks of JI, one needs to know more about abatement costs, especially in developing countries.

2 The UNEP Abatement Costing Project

A great deal of work has been done and more is underway to determine emission inventories and establish emission scenarios for developing countries. For example, a series of country studies has been coordinated by Lawrence Berkeley Laboratory (LBL), and these studies include emission scenarios and identify reduction measures.[1] Although subsequent studies have begun to address costs, the existing

---

* The authors are associated with UNEP Collaborating Centre on Energy and Environment Risø National Laboratory, Denmark.

1 Sathaye and Ketoff, 1991.

C. J. Jepma (ed.), The Feasibility of Joint Implementation, 249–266.

studies do not analyze costs directly, and indeed there are few studies that have estimated emission abatement costs at the national level for developing countries.

The most ambitious abatement costing effort to date is the UNEP Greenhouse Gas Abatement Costing Study, which established and tested a methodological framework for national studies of the costs of reducing emissions and applied it in Brazil, Egypt, India, Senegal, Thailand, Venezuela, and Zimbabwe as well as in Denmark, France and The Netherlands.[2] Thus, this project had a major emphasis on developing countries.

2.1 Methodology

The methodological framework for GHG abatement costing comprises a consistent set of basic concepts and definitions, such as prices and costs, input data assumptions and a uniform analytical structure.[3] This framework comprises three elements:

- definition of key concepts and terminology;
- a set of common quantitative assumptions (international fuel prices, test discount rates, reduction targets, *etc.*);
- an analysis structure (see Figure 18.1):
    - country review (background and existing studies);
    - definition of a reference scenario;
    - identification and ranking of abatement options;
    - construction of abatement scenarios;
    - macroeconomic assessment; and
    - subsequent evaluation (political, social, *etc.*).

The abatement costs are measured as direct/financial costs for individual technical options and on a sectoral level for the energy system in total. The approach of the country studies was 'bottom-up' technical-economic investigations of the reference case and abatement scenarios. Following this approach, most of the national teams were able to construct consistent abatement scenarios for the specified targets of 12.5%, 25% and 50% reduction from baseline scenario emissions. India, Thailand and Venezuela stopped at a reduction of less than 30% from the baseline in 2030 as

---

2 UNEP, 1992, 1993, 1994.

3 UNEP, 1993.

the maximum reduction.[4] Reduction targets are based on reference scenario emissions, not current emissions.

Figure 18.1. Structure of proposed abatement analysis procedure.

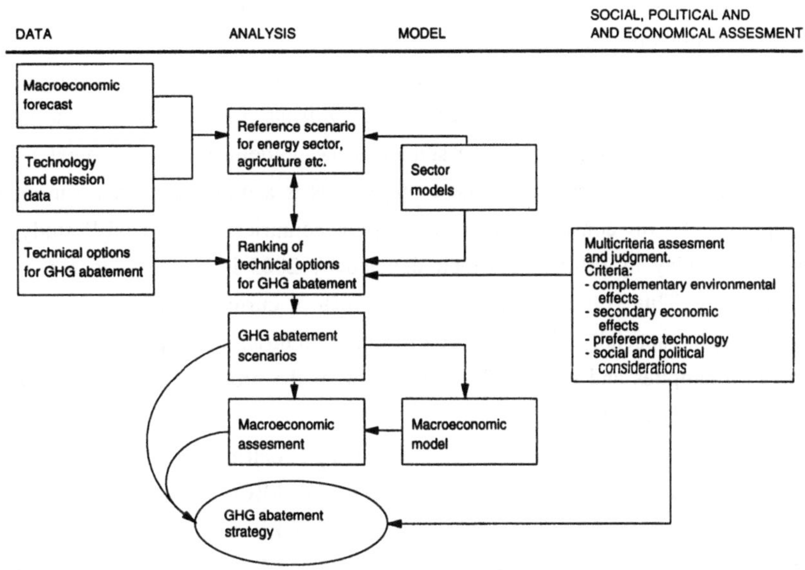

2.2 *Reference and abatement scenarios*

A national GHG abatement costing study is dependent on the definition and construction of scenarios. The cost of GHG abatement is defined in terms of the difference between a *Reference Scenario* and an *Abatement Scenario*. The term scenario is taken to mean a consistent description of the country-in-question, particularly its energy system and GHG emissions, over the period of the study, from the present to 2025 or 2030.[5] Thus, a scenario consists of a set of assumptions concerning national economic and demographic development, technical details of energy consumption and supply systems, for the country in question, and for the international energy economy, to the extent that this affects GHG emissions and abatement costs in the country.

[4] UNEP, 1994.

[5] UNEP, 1993.

Because the abatement scenarios are defined in terms of reductions from a baseline scenario, the emission results depend on the definition of the baseline. The results are sensitive to the degree to which potential low-cost energy-efficiency improvements are captured in the reference scenario, rather than being available as emission-reduction measures in the abatement scenarios. The abatement cost results are especially sensitive to the definition of the baseline scenario and the energy-efficiency assumptions used.

The reference scenarios tend to be influenced heavily by recent developments, consistent with the assumption that cheap energy will last for many years, due in part to the global GHG abatement effort assumed in the guidelines as the context for the national studies. A consequence of these projections is a dramatic increase in energy consumption. The oil consumption of developing countries in the UNEP study was 3.5 mln. barrel/day in 1990, while their demand by 2025 is projected to reach 16.2 mln. barrel/day. This volume represents about 70% of current OPEC production and is large enough, especially if extrapolated to other countries, to impose great pressure on world oil supplies and prices. The consistency of the projected oil prices is thus a matter for some discussion.

The reference scenarios for the seven developing countries calculate that the total CO_2 emissions from the group will increase from 750 $MtCO_2$ (200 MtC) in the base year to about 3500 $MtCO_2$ (950 MtC) by the end of the period in 2025-2030. The abatement scenario reduces CO_2 emissions to 2100 $MtCO_2$ (570 MtC). This corresponds to a 40% reduction in emissions from baseline, but still almost a tripling of emissions compared to 1990 levels. The full 50% reduction target was not achieved in the aggregate because a few of the countries did not achieve feasible solutions at the 50% reduction level (India), or because all abatement options were not exhausted (Thailand and Venezuela).

The country studies show substantial differences in terms of the amount of CO_2 emissions per capita (Figure 18.2). For example, the Thailand reference scenario projects a six-fold increase in per capita emissions, while in Senegal and Zimbabwe per capita emissions increase very slowly. Comparison in absolute terms reveals a huge gap among the countries: CO_2 emission levels per capita in Thailand are expected to exceed by more than 20-fold those of India and reach the levels of the industrialized countries by 2030.

Figure 18.2. Per capita CO_2 emissions of the participating countries in the base year and end years of reference and abatement scenarios.

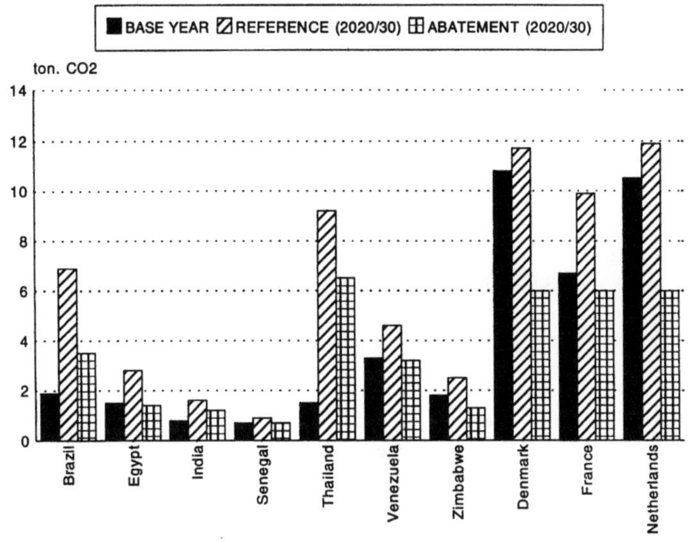

3 Cost Results from the UNEP Country Studies

Abatement cost is defined as the incremental cost of GHG reductions, relative to a baseline in which abatement is not an objective.[6] Costs are measured as the direct/financial costs on an energy system and sectoral level. The costs include levelized investment, operation and maintenance and fuel costs. Thus, the emission reductions and costs are defined in terms of differences from the reference point. The incremental cost assessment is therefore closely interrelated with the definition of the baseline.

Cost curves, containing marginal abatement costs for several target reductions in a specific year, illustrate how the cost varies with the degree of reduction and allow different country studies to be compared. Figures 18.3 and 18.4 show marginal abatement costs for most of the participating countries in the short and long term, respectively.[7] The cost curves for developing countries exhibit a number of similarities.

[6] UNEP, 1993.

[7] UNEP, 1994.

Figure 18.3. Marginal abatement costs for emission reductions in the short-term target year ($/t$CO_2$).

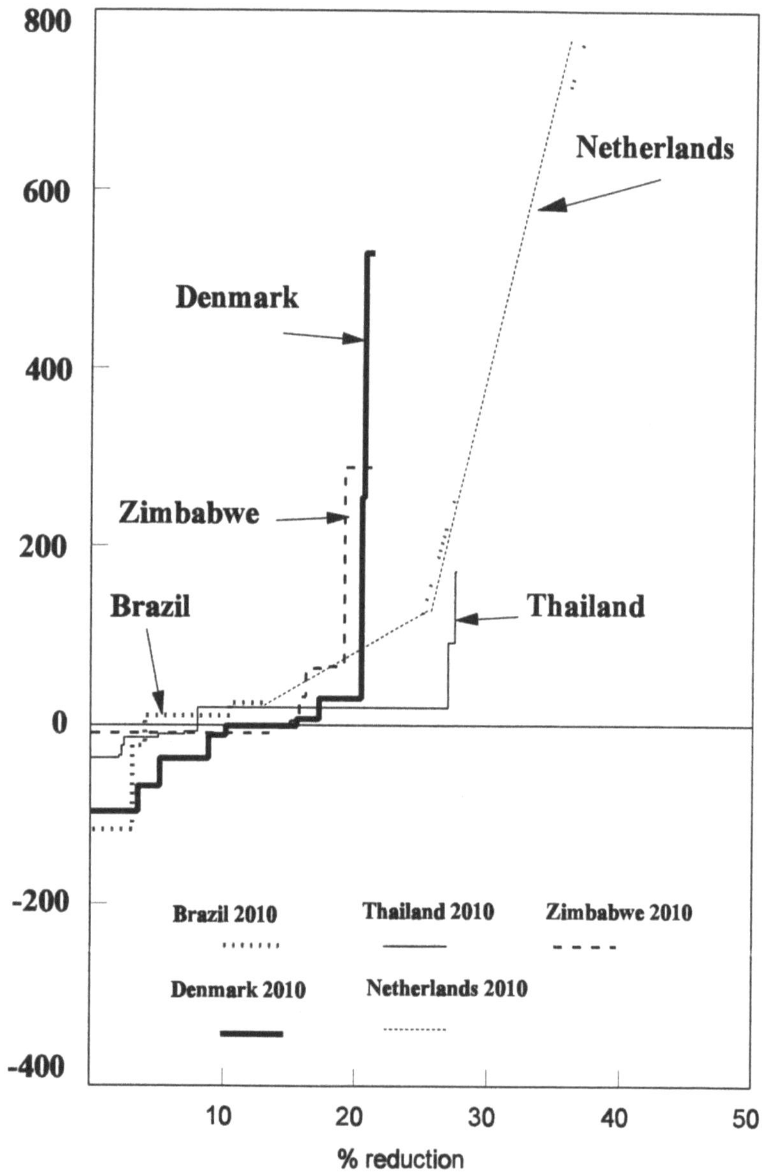

Figure 18.4. **Marginal costs for emission reductions in the long-term target year ($/tCO_2$).**

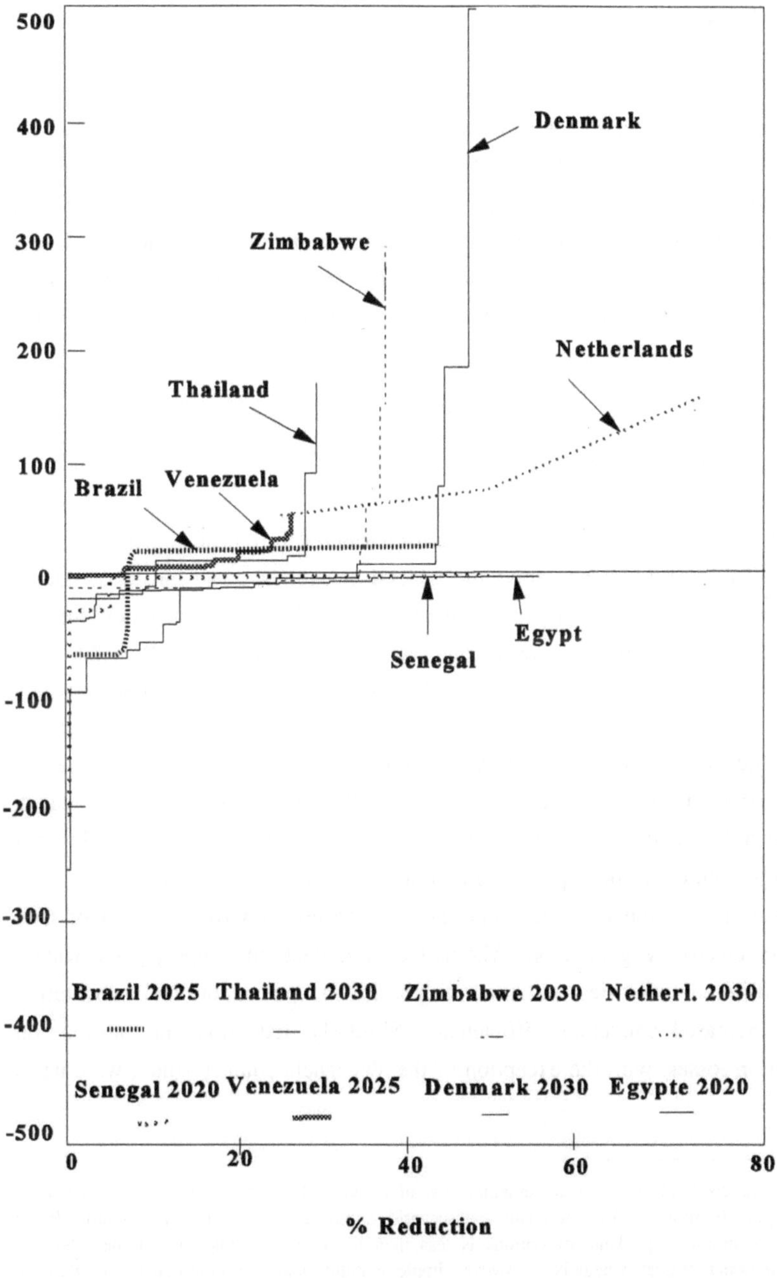

A particular feature is the large potential for negative cost abatement options,[8] amounting to up to 20% reductions from the baseline in 2005-2010 and about 35% in 2020-2030. The curves are also similar in shape: the first part of the cost curve, up to 5 to 10% emission reductions, indicates very low abatement cost, followed by a long interval, up to about 25% reduction in the short term and about 40% in the long term, in which marginal abatement cost falls within a range between -10 and +30 $/t$CO_2$ (40-110 $/tC) reduced.

The abatement cost curves develop in a similar way as the time horizon for emission reductions is extended. Over time, the cost curves become less steep and move downward in absolute cost. This reflects the fact that, over time, increasing opportunities for capital replacement allow the use of more efficient and less polluting energy technologies in new and replacement applications. Moreover, these technologies tend to be more effective and less expensive than retrofits of existing equipment.

4 Abatement Technology Options

The general classes of technologies that were considered in the abatement scenarios are listed in Table 18.1, which shows the general categories of abatement measures that provided significant emission reductions in the abatement scenarios for the different countries. An important result that can be seen from the table is the wide variety of end-use measures considered in the country studies. The category which was examined most intensively by the countries was industrial energy efficiency, the details of which depend on the industrial structure of the countries. The residential sector was covered more generally by most of the developing countries, which assessed either fuel substitution or energy-efficiency options for that sector, but not both, although both could be expected to be relevant for all the countries.

Most of the supply-side abatement options were focused on the potential of CO_2 reductions in electricity generation. Alternatives for electricity supply vary widely among the different countries, but most emphasis was placed on fuel substitution rather than increased conversion efficiencies. Similarly, reductions in transmission and distribution losses, with the exception of the Venezuela country study, were not a

---

[8] 'Negative cost' refers to the incremental cost of meeting the demand for equivalent energy services (transport, lighting, etc.) in the abatement scenario, compared to the reference scenario. If the *total* cost of an option or package of options is less than the corresponding cost in the reference scenario, the *incremental* cost is negative. However, implementation and transaction costs are often not included in the direct cost calculations in the scenarios.

priority area. Direct use of solar or wind power is not a significant factor in most of the abatement scenarios. Large-scale contributions from renewable energy sources to electricity generation were considered only in Denmark, with increased use of wind power, and Egypt, where wind and photovoltaic systems are projected to contribute 19% of the electricity supply in the abatement scenario. Both of these countries already have significant reliance on the same renewable sources in their reference scenarios.

Table 18.1. Abatement options applied in the country studies.

Abatement option	Br	DK	Eg	F	Ind	NL[1]	Sen	T	V	Zw
Energy use										
Agricultural fossil fuel saving			X		X					X
Building energy efficiency	X	X		X[2]		X		X		
Building fuel substitution		X	X		X	X	X			
Industrial cogeneration			X		X	X				
Industrial energy efficiency	X	X	X	X[2]		X	X		X	X
Industrial fuel substitution	X	X	X		X	X				
Solar heating		X	X			X				
Transport fuel substitutions	X		X		X	X				
Transport modal shifts			X							
Vehicle energy efficiency			X			X		X	X	
Energy supply										
Biomass fuel	X	X				X				X
Biomass power	X	X				X				X
Hydropower			X							X
Natural gas						X			X	
Nuclear power				X[2]				X		
Reduced power losses									X	
Solar power			X			X				
Wind power		X	X			X				
Forestry		X		X[2]						

1 The Netherlands' study includes in addition CO_2 removal technologies and the use of hydrogen as an energy carrier.

2 Individual technical measures are not listed in the French report.

While the range of options considered in the scenarios is an interesting result of the country studies, one might also consider the options that could be assessed further in the country studies, if more information, time and resources were available. Some of the countries omitted some possible abatement options particularly for far-reaching emission reductions above 20% reduction in the short term and 35% in the long term. For example, if energy-efficient lighting is cost-effective in Brazil, this result should also apply to Venezuela. Also, supply-side options such as reduced transmission losses, biomass-fired cogeneration and wind power may be applicable to several countries other than the few that considered these options. Although the economies of energy efficiency improvements and renewable sources are highly site-specific, it is unlikely that they vary so much from country to country as to completely exclude these options from the ranking of options.

Other aspects of the analysis, however, may lean in the optimistic direction. For example, the market penetration for certain energy-efficient technologies appears to be near 100% in the abatement scenarios for Denmark, Egypt and Zimbabwe. This kind of assumption may be unrealistic, but it may be balanced by the omission of other abatement options in the same studies. All in all, we would conclude that the approach of the country studies is reasonably realistic, and slightly pessimistic with regard to technology assumptions.

A major discussion point in the development of the guidelines was whether a standard set of technology data should be recommended, in order to identify abatement measures and analyze their costs and performance. This was rejected, because a high priority was given to the evaluation of the appropriateness and performance of different technologies in the national context. Instead, this information was developed within each participating country. The results of the country studies, with regard to both the low-cost abatement potential and the omission of additional potential options, suggest the need for more international effort to identify and evaluate energy efficiency and emission abatement options, especially within and among developing countries.

5 Implementation Issues

One general similarity among all the studies is that the least expensive part of the cost curve contains energy end-use savings in industry and/or households. For the developing country studies in general, the profitability and possible penetration rates of such energy-saving end-use technologies are closely linked to general macroeconomic efficiency improvements and possible energy price adjustments to world market prices. In many cases the removal of general inefficiencies in the economic system could be combined with an abatement policy. However, the implementation of both efficiency improvements and pure abatement options may be hindered by institutional barriers.

Some investments can contribute indirectly to emission abatement through overall economic efficiency gains, and must therefore be made in advance of investments directly aimed at emission abatement. In addition, some aspects of the implementation process, even for cost-effective measures such as energy-efficiency improvements, will require investments in information programmes, demonstration projects, testing facilities, training and outreach, and general capacity building.

Although the net cost of a given level of resulting emission abatement may be small or negative, the necessary implementation costs appear as positive cost barriers, because it is difficult to allocate these costs to individual abatement measures and to the stake-holders in the measures.

Thus, overcoming the implementation barriers to energy-efficiency improvements and other potential low-cost options requires institutional changes and involves transaction costs. Such implementation issues and costs were generally not treated explicitly in the UNEP country studies. Nor is it yet possible to model synergistic benefits from introducing information, technological learning, removing risk through comprehensive regulations and incentives, and reduced emissions of non-GHG local pollution.

If these countries are to carry out the types of abatement measures identified, however, such implementation costs and external benefits will become important considerations in their strategy. Investments in institutional capacity-building and other implementation steps will facilitate additional investments in abatement measures that would be considered cost-effective. Thus, the initial investment elements should be considered incremental and necessary for protecting the global environment and therefore eligible for international funding from, for example, the Global Environment Facility (GEF).

6 Forestry Sector Studies

The UNEP studies did not cover the forestry sector. However, because tropical deforestation is a major source of GHG emissions, the possibility of reducing or reversing this process, through protection of existing forested lands, reforestation and afforestation, is part of a global abatement strategy, and possibly a relatively inexpensive component. Also, the prospect of using harvested biomass as an energy source to replace fossil fuels makes the potential for preventing carbon emissions even larger over the long term.

As in the case of energy-sector analysis, most of the national-level studies done to date have addressed emission inventories and scenarios, including a series of country studies coordinated by LBL and some continental-scale analyses partially based on satellite imagery.[9] A partial assessment of national costs of carbon storage via reforestation, based on site-level data, was made for several of the large forested countries.[10] For the tropical countries considered, the costs varied from \$1/tCO_2 (\$4/tC) for Mexico and Malaysia, to \$3/tCO_2 (\$10/tC) for Brazil, China and South Africa, up to \$4-5/tCO_2 (\$15-18/tC) for India and Argentina.[11] These costs may be incomplete, as they do not include opportunity costs of land or maintenance and management costs. Moreover, they include only reforestation projects.

Another set of bottom-up case studies analyses the full net costs of carbon storage through reforestation and conservation projects in Central America.[12] The approximate costs of the projects studied in Costa Rica range from \$1/tCO_2 (\$4/tC) for forest reserves to \$4/tCO_2 (\$15/tC) for small-holder woodlots and \$7/tCO_2 (\$25/tC) for restoration of degraded natural forest ecosystems. Approximate costs of reforestation are \$3/tCO_2 (\$10/tC) in Panama and in Guatemala range from \$4/tCO_2 (\$15/tC) for timber plantations to as low as \$1/tCO_2 (\$4/tC) for rural agroforestry programmes. Other project-level analyses have found similar carbon storage potential for reforestation with somewhat lower costs of \$2-3/tCO_2 (\$8-10/tC) in Ecuador and Brazil.[13] The carbon storage values for plantation projects are mostly between 300 and 400 tCO_2/ha (80-100 tC/ha), depending on climate and project design.

When the same projects are considered as biomass energy plantations, the costs per ton are similar, and the carbon savings per ha generally increase 20-60%, depending on the conversion technology, over the 40-year life of a power station project, or

---

9 Makundi et al., 1992; Houghton et al., 1991.

10 Winjum et al., 1993.

11 These are undiscounted cost values. The costs are generally incurred when a project is initiated, and the net carbon storage can occur almost immediately, gradually over the project life, or decades later, depending on the type of baseline or reference land use. It is problematic to attempt to discount carbon, which may have a different time-value than the monetary time-value indicated by the discount rate, so the carbon storage values represent a one-time carbon removal or prevention of carbon emissions, which is assumed permanent. These cost values are not exactly comparable to costs of energy-sector measures to prevent annual carbon emissions.

12 Swisher, 1991.

13 Swisher, 1991; Winjum and Lewis, 1993.

about 3-6 harvest rotations.[14] With biomass energy plantations, however, additional power stations could be built and operated in the future, accumulating carbon savings beyond the one-time values calculated above, provided that biomass fuel can continue to be harvested sustainably.[15]

The carbon-saving results for the different types of projects were used to estimate national carbon-storage potential for Costa Rica and other Central American countries, based on project proposals and cost estimates given in the national Tropical Forest Action Plan (TFAP). The national TFAP is not meant to suggest the total potential for forestry investments in a country, nor does it specifically consider carbon storage (FAO, 1986). It does, however, provide a list of projects that could be carried out if sufficient funding were available for the forest sector. This list can be used to construct a national-level estimate of carbon-saving potential and cost. In addition to the TFAP projects, further timber and fuelwood plantations can be considered, with similar project costs, but with increasing opportunity cost for the land, based on land-value data for rural land with minimum services.

Total carbon storage potential in Costa Rica, at a marginal cost below \$3/tCO_2 (\$10/tC), is at least 400 MtCO_2 (110 MtC) in Costa Rica and nearly as much in other Central American countries.[16] Most of the carbon savings from the plan come from protection of standing biomass on forest lands vulnerable to deforestation, while about 15% comes from new biomass in forest restoration, natural forest management, agroforestry and plantation projects. Including reforestation on more expensive land than that considered in the TFAP project list increases the total carbon storage potential to over 500 MtCO_2 (140 MtC), but at a marginal cost of about \$8/tCO_2 (\$30/tC).

Additional forest-sector analyses have been done for Brazil, the country with the largest area of tropical forests and recently the highest rate of deforestation. One of the LBL country studies estimates the present rate of emissions from deforestation at 1,000 MtCO_2 (280 MtC/year); the net cost of preventing the land-use change in the Amazon region would be negative, while other estimates give a maximum of \$1/tCO_2 (\$4/tC) as the net cost.[17] The proposed Floram project in Brazil would provide for the reforestation of 20 Mha of degraded land in southern Brazil, over the next 30 years, and the restoration of 45 Mha of savannah in northeastern Brazil, all outside

[14] Swisher, 1994.

[15] Marland and Marland, 1992.

[16] Swisher, 1992.

[17] Fearnside, 1992; Moreira et al., 1992.

the Amazon region. The estimated carbon storage is 13 bln. tCO_2 (3.6 bln. tC) from reforestation and 5 bln. tCO_2 (1.4 bln. tC) from savannah restoration, at an approximate cost of $22.5 bln. or an average cost of $1/$tCO_2$ ($4/tC).[18] It appears, however, that the carbon storage is exaggerated by as much as a factor of two, so the average cost may be closer to $2/$tCO_2$ ($8/tC).

The low costs of carbon storage indicate that forestry projects would be attractive for JI. This strategy has been criticized as reserving land in developing countries to clean up industrialized countries' waste. However, the premise of carbon offsets is not that land is freely available in developing countries to displace other land uses and plant trees of any type desired. Rather it is that a real need exists for reforestation, forest management and biomass energy development, which will not be financed under existing conditions, but that offer both local and global benefits if implemented under the proper conditions.

7 Interpretation of the UNEP Abatement Costing Results

One can see from the cost curves in Figures 18.3 and 18.4 that developing countries appear to have significant quantities of low-cost, and perhaps negative-cost, options for reducing emissions from the energy sector. This result validates one of the principal assumptions behind JI, that the least-cost emission abatement strategy includes measures in developing countries not bound by existing reduction commitments under the FCCC. In addition, there are low-cost opportunities for carbon storage in the forestry sector, especially in Latin America. However, this abatement option is a one-time carbon sink, rather than an annual emission reduction option as in the energy sector.

It is important to observe that there also appear to be significant low-cost abatement options in the industrialized countries, especially in a case like Denmark, with a large negative-cost abatement potential. This result is consistent with other studies of industrialized countries.[19] It is sometimes assumed that interest in JI is based on a lack of cost-effective abatement options in industrialized countries.[20] However, the percentage reductions possible at relatively low costs appear comparable for industrialized and developing countries. In the short term, absolute emissions will continue

---

[18] Instituto de Estudos Avançados, 1990.

[19] Blok, 1991; Jackson, 1991; Swisher and Johnson, 1992; Krause et al., 1993; Lazarus et al., 1993.

[20] Loske and Oberthür, 1994.

to be much greater from industrialized than from developing countries, and thus the abatement potential of the former will also be larger during the next decades.

7.1 A tentative extrapolation

If one takes the countries considered in the UNEP GHG Abatement Costing Study as representative of the global population of industrialized and developing countries, it is possible to estimate the conditions for JI in the abatement scenarios from these country studies. Assume that the industrialized countries are committed to an emission reduction path that follows the abatement scenarios, which corresponds to an annual reduction of about 1% in absolute emissions, countries' emissions are not constrained. In the short term, until 2005-2010, the industrialized countries could exchange at most about one-quarter of their 25% reduction from baseline emissions via JI with the developing countries. In such a case, the 'market clearing' price for carbon offsets would be more than \$30/tCO_2 (\$110/tC), assuming no transaction costs. Without JI, the *marginal* abatement cost could be as high as \$110/tCO_2 (\$400/tC). With JI, the *average* abatement cost would be reduced from about \$25/tCO_2 (\$90/tC) to about \$10/tCO_2 (\$40/tC), assuming no transaction costs.

The remaining (approximately) 18% emission reductions would be less expensive to implement domestically, or through JI between industrialized countries, than through additional JI with developing countries. Some additional low-cost offsets could be obtained through forestry measures. However, the implementation barriers discussed earlier would tend to reduce the size of emission abatement potential at a given cost level, especially in developing countries, and transaction costs would increase the cost levels. Also, in the early years around 2000, the relative abatement potential in developing countries would be less than in the later years of the scenario.

While JI cannot remove industrialized countries domestic emission reduction commitments, it would appear that JI could remove the most expensive and therefore painful reduction obligations and reduce the average cost of a given level of abatement. Alternatively, larger emission reductions could be achieved at a given total cost level. Assuming that implementation and transaction costs are insignificant, the absolute scale of the abatement options captured through JI would still be relatively modest. Moreover, the marginal costs of domestic emission reductions would certainly be high enough to provide serious incentives for technical and institutional innovations to reduce domestic costs in the industrialized countries.

7.2 The issue of timing

The rationale of economic efficiency is that total costs should be minimized by implementing the least expensive options available, which might be expressed as 'cheapest measures first', in priority, but not necessarily in time. This idea is sometimes interpreted to mean that 'the cheapest possibilities for emission abatement will have been used up by the industrialized countries'.[21] In practice, however, one has little control over the timing of most cost-effective abatement measures.

Preventing deforestation, for example, is an option that only exists as long as the forest remains. Similarly, inexpensive energy-efficiency opportunities appear at the time that buildings and equipment are constructed, when the net cost of efficiency improvements are small, and these opportunities disappear or become much more expensive as soon as the buildings or equipment enter service. Decisions about what kind of power station to build are also one-time opportunities. Once the forest has fallen or the inefficient equipment is installed, it is either impossible or very costly to reverse the process. If a strategy of 'cheapest measures first' is not followed, such as via JI, it will not mean that these measures can be stored away to be captured later; rather it means that they will be missed entirely. Thus, it is an advantage of JI, not a disadvantage, that it can help capture inexpensive abatement opportunities in developing countries sooner rather than later.

---

21 Loske and Oberthür, 1994, p. 53.

References

Blok, K., 1991. *On the Reduction of Carbon Dioxide Emissions*, PhD Thesis, Univ. of Utrecht.

Fearnside, P.M., 1992. 'Carbon Emissions and Sequestration in Forests: Brazil', Lawrence Berkeley Laboratory, LBL-32758.

Food and Agriculture Organization (FAO), 1986. *Tropical Forestry Action Plan*. FAO, Rome.

Hayes, P., 1993. 'North-South Transfer', in: Hayes, P. and K. Smith (eds.), *The Global Greenhouse Regime: Who Pays?*' Earthscan, London.

Houghton, R.A., J. Unruh, and P.A. Lefebvre, 1991. *Current Land Use in the Tropics and its Potential for Sequestering Carbon*, Proceedings, Technical Workshop to Explore Options for Global Forestry Management, pp. 297-310, Bangkok.

Instituto de Estudos Avançados, 1990. 'Projeto FLORAM - uma Platforma', *Estudos Avançados,* Universidade de São Paulo, vol. 4, no. 9.

Jackson, T.,1991. 'Least Cost Greenhouse Planning', *Energy Policy*, vol 19, no. 1, pp. 35-46.

Jones, T., 1993. *Operational Criteria for Joint Implementation*, Proceedings, International Conference on the Economics of Climate Change, OECD, Paris, June.

King, K., 1993. *The Incremental Costs of Global Environmental Benefits*, Global Environment Facility, Working Paper Number 5, World Bank, Washington.

Krause, F., E. Haites, R. Howarth, and J. Koomey, 1993. *Cutting Carbon Emissions: Burden or Benefit*, International Project for Sustainable Energy Paths, El Cerrito, Calif.

Lazarus, M., L. Greber, and J. Hall, 1993. *Towards a Fossil Free Energy Future*, Stockholm Environment Institute Boston Centre.

Loske, R. and S. Oberthür, 1994. 'Joint Implementation under the Climate Change Convention', *International Environmental Affairs*, vol. 6 no. 1, pp. 45-58.

Makundi, W., J. Sathaye, and O. Masera C., 1992. *Carbon Emissions and Sequestration in Forests: Case Studies from Seven Developing Countries*, Lawrence Berkeley Laboratory, LBL-32119.

Marland, G., and S. Marland, 1992. 'Should We Store Carbon in Trees', *Water, Air and Soil Pollution*, vol. 64, pp. 181-95.

Merkus, H., 1992. *The Framework Convention on Climate Change: Some Thoughts on Joint Implementation*, Ministry of Housing, Physical Planning and the Environment, Den Haag, The Netherlands, CCD/Paper 11.

Moreira, J.R., G.M. Jannuzzi, and A.D. Poole, 1992. 'Abatement of CO_2 in Brazil', in: R.K. Pachauri and P. Bhandari (eds.), *Global Warming: Collaborative Strategies to Limit Carbon Dioxide Emissions in Asia and Brazil*, Tata McGraw Hill, New Delhi.

Sathaye, J., and A. Ketoff, 1991. *CO_2 Emissions from Developing Countries: Better Understanding the Role of Energy in the Long Term*, Lawrence Berkeley Laboratory, LBL-29507.

Smith, K., 1991. 'Allocating Responsibility for Global Warming: The Natural Debt Index', *Ambio*, vol. 20, pp. 95-6.

Smith, K., J. Swisher, R. Kanter, and D. Ahuja, 1993. 'Indices for a Greenhouse Gas Control Regime: Incorporating Both Efficiency and Equity Goals', in: Siddayao, C. (ed.), *Energy Investments and the Envrionment*, World Bank, Washington, DC.

Swisher, J. and K. Johnson, 1992. 'Cost and Performance Comparison of DSM Measures and Supply Changes to Reduce Utility Carbon Emissions', Proceedings, *ACEEE Summer Study on Energy Efficiency in Buildings*, Monterrey, Calif., September.

Swisher, J. and G. Masters, 1989. *International Carbon Emission Offsets*, Stanford Univ., Civil Engineering Dept. Technical Report TR-309, Palo, Alto, Calif.

Swisher J. and G. Masters, 1992 'A Mechanism to Reconcile Equity and Efficiency in Global Climate Protection', *Ambio,* vol. 21, pp. 154-9.

Swisher, J., 1991. *Prospects for International Trade in Environmental Services: An Analysis of International Carbon Emission Offsets*, PhD Thesis, Stanford Univ., Civil Engineering Dept.

Swisher, J.N., 1992. 'Cost and Performance of CO_2 Storage in Forestry Projects', *Biomass and Bioenergy*, vol. 1, no. 6, pp. 317-28.

Swisher, J.N., 1994. 'Forestry and Biomass Energy Projects: Bottom-Up Comparisons of CO_2 Storage and Costs', *Biomass and Bioenergy*.

UNEP, 1992. *UNEP Greenhouse Gas Abatement Costing Studies. Phase One Report.* UNEP Collaborating Centre on Energy and Environment, Risø, Denmark.

UNEP, 1993. *Methodological Framework for National Greenhouse Gas Abatement Costing Studies: Phase II Guidelines.* UNEP Collaborating Centre on Energy and Environment, Risø, Denmark.

UNEP, 1994. *UNEP Greenhouse Gas Abatement Costing Studies. Phase Two Report.* UNEP Collaborating Centre on Energy and Environment, Risø, Denmark.

Winjum, J.K., R.D. Dixon, and P.E. Schroeder, 1993. 'Forest Management and Carbon Storage: An Analysis of 12 Key Forest Nations', *Water, Air and Soil Pollution*, vol. 70, pp. 239-57.

Winjum, J.K., and D.K. Lewis, 1993. 'Forest Management and the Economics of Carbon Storage: the Non-Financial Component', *Climate Research*, vol. 3, pp. 111-9.

19. Energy Savings Potentials, Issues and Constraints

MOHAMMED ASADUZZAMAN*

1 Introduction

Since the first oil shock in the 1970s, energy conservation has become the catch word in many countries. Such attempts received a boost in the wake of the recent scientific findings and awareness about global climate change and its possible and profound societal and other impacts. As is well-known by this time, the global warming and the associated global climate change are brought about by the atmospheric accumulation of certain gases like CO_2, mainly caused by fossil fuel burning and other activities associated with energy use. As a result, we now hear about various policy measures like carbon tax, tradeable permits in emissions, Joint Implementation (JI) and the like. The purpose of this chapter is to review potentials of energy savings and the issues and constraints that arise in realizing those potentials. The focus will mainly be on economic and social issues.

Section 2 will discuss the concepts of energy saving and energy intensity while a brief idea about the technical potentials of energy savings will be given in section 3. Sections 4 will focus on the experiences of various countries with energy conservation and efficiency and the constraints faced in putting energy conservation policies into effect. The concluding section 5 will try to draw some lessons for future policy formulation.

2 Energy Use, Savings and Efficiency: Some Conceptual Clarifications

Energy is used for various purposes in production, distribution and consumption activities. One obvious implication is that for comparison purposes (whether across nations or between sectors within a country), some form of common denominator (like heating value or some energy commodity equivalent itself being based on the energy content, *e.g.* tons of oil equivalent) has to be used.

---

* The author is a senior research fellow in the Bangladesh Institute of Development Studies. The views expressed here are the author's own and draw heavily upon his recent research (on behalf of IPCC) on energy efficiency. These views are not necessarily shared by his employer Institute.

C. J. Jepma (ed.), The Feasibility of Joint Implementation, 267–284.

Absolute vs. relative use of energy

The next issue is how to express energy use, as an absolute value or a relative value? It is easy to show that the two are linked through some accounting balances (see Annex). As lowering total energy use is the ultimate goal, particularly if looked at from the angle of the CO_2 emissions, we can use this accounting balance equation to focus on the areas for policy action over different lengths of time period: short, medium or long.

There are two relative energy use indicators which are of interest here. One is the energy use per unit of GDP (energy intensity of GDP) while the other is the per capita use of energy. It probably goes without saying that energy conservation should mean a fall in energy intensity of GDP. But whether or not per capita energy use should fall remains a debatable issue and depends on the initial condition of the particular country concerned. Looking at equation (3) in the Annex, it can be seen that the change in total energy use directly varies with per capita use of energy. Hence, the scope for a reduction in total energy use is greater, to begin with, in countries with very high per capita energy use (*i.e.* in the industrialized countries). A similar conclusion arises regarding the relative roles of large and small economies. The initial scope for energy savings is greater in the case of the former. As the high per capita energy-using countries in general also have large economies (compared to Third World countries of similar-sized population), it is clear that the major thrust for energy conservation, at least for the time being, should be in the Western countries and the countries in Central and Eastern Europe and the CIS.

Developing countries with smaller economies and low per capita energy use will, of course, also have to share the burden and increasingly so as their economies begin to resemble more and more those of the industrialized countries. If one is allowed to view the issue of per capita energy use as an indicator for equity and that of energy intensity as one for efficiency, a global arrangement among nations may take care of the problem of ensuring both efficiency and equity in energy conservation at the same time. This immediately raises the problem of defining 'energy efficiency'.

Energy efficiency

According to economists, energy is used most efficiently (in production) if the value of its marginal product equals the price paid for it. Given two different prices of energy, therefore, two firms in two different situations may both be using energy most efficiently, yet the energy intensity of the output they produce may be different and in the energy literature the one with lower energy intensity is termed to be more energy efficient. The general idea of efficiency in the energy literature is therefore rather akin to managerial or *X*-efficiency. This is the crux of the problem. While policy makers may want to lower energy intensity and thereby save energy, the firms may under normal circumstances see no reason why they should do so, since their main goal is to strive for economic and *X*-efficiency.

Sectoral comparisons: direct and indirect use of energy

An important issue that arises relates to the comparison between sectors, with respect to the problem of whether energy is used directly or indirectly. Direct use refers to the consumption of energy in the production process of a good or service. Indirect use refers to the energy expended in production of goods and services which are used as intermediate inputs in the production process of the first good or service. It may be that the rankings of sectors by energy intensity will be different when, on the one hand, only the direct energy use is considered, and when, on the other hand, both the direct and indirect uses are taken into consideration, which will consequently have different policy connotations. Unfortunately, one rarely finds awareness of this issue in the literature.[1]

Structural shift in economy and energy intensity

Another issue relates to the structural shift in the economy and is thus linked to the above mentioned inter-industry flows. If, over time, the economy produces relatively more goods and services which demand less energy per unit of output, then the energy intensity of GDP may fall without any rise in the energy efficiency in the production of individual goods and services. Energy intensity may therefore be divided into: the structural effect, without any change in energy use per unit of value of the individual goods and services produced; and a pure intensity which measures the energy expended per unit of value created in the production of those individual goods and services. It is this effect that will be referred to as energy use efficiency with the highest potential for energy savings. The more efficient a (production, distribution or consumption) system is, the more energy it saves, at least in a relative, if not in an absolute sense. If the rate of efficiency gain (whether through technological change or other means) is greater than the rate of change in the size of the energy-using system, there will also be an absolute fall in energy use. Ensuring such an outcome, though not impossible, may not be easy as it would depend on many factors (see sections 4 and 5), some of which (like population) have rather long-run dynamic effects.

---

[1] See, however, Parikh and Gokarn, 1993; and Common and Salma, 1992.

3 Technical Potentials for Energy Savings

Energy saving in the industrialized countries

Energy efficiency gains may be made in the following four areas (Wilson, 1988): retaining useful energy in the product; motive power; process heat; and space and water heating. The theoretical scopes for such gains are enormous. Table 19.1 illustrates such possibilities even with the available, existing technologies. Electrical motors which mostly supply industrial motive power can utilize only 35% of the electricity applied to them.[2] In Japan, an estimated one-half of the total energy produced was wasted even during the mid-1980s[3] In USA, the potential savings in residential appliances were estimated to be 50-87%.[4] More recent estimates for the USA show energy saving potentials of 45% in buildings and 30% in industries.[5] For the world as a whole estimates show that a 30% energy conservation may be possible by the year 2030, 30% in the case of manufacturing and transportation and 50% in the case of electricity generation/transmission and residential use.[6]

Table 19.1. Illustrative Indicators of Energy Saving Potentials.

Energy using activity	Energy intensity measure	Present average	Alternative energy intensity	
			Present best technology	Advanced technology
automobile	litres gasoline eq. /100 Km	10 -15	5	2.3 - 3.6
residential space heating	kJ of heat output/m²/degree/day	120 - 160	50	15 - 18
refrigeration	kWh/litre/year	3.4	1.3 - 1.7	0.7
steel production	GJ of final energy/mt of raw steel	22 - 27	15	9 - 12

Source: Goldemberg, et al., 1988, p. 87.

2 Wilson, 1980.

3 Belgrave et al., 1987.

4 Flavin and Durning, 1988.

5 Rubin et al., 1992.

6 ESCAP, 1991, Table 5.3.

Energy savings in the developing countries

The issues of energy conservation and efficiency in the developing countries are, despite the initial major role to be played by highly industrialized nations as suggested earlier, very important but also somewhat different in nature compared to those in the latter. In developing countries, a very large part of the demand for energy is met from renewable energy sources like biomass.

At the same time the demand for fossil fuels and electricity, however, is quickly increasing while the efficiency in their use, particularly in industrial activities, have shown little or no improvement.[7] Despite the general lack or inadequacy of funds for the development or imports of energy resources in these countries and the competition among different sectors for the same funds, there are countries which have to allocate a quarter to a third of public funds to power generation. Even these large relative shares are inadequate to meet the growing demand.[8]

It may also be pointed out that while the present industrialized nations are experiencing a structural shift in their economies from high energy-intensive secondary to a less energy-intensive tertiary sector, the developing countries in general are entering more and more into the secondary sector. The imperatives for energy conservation and gains in energy efficiency are of paramount importance in such a situation. Because of the fast growth in commercial energy use, the increasing role of industrialization and the current low efficiency level, the potentials for energy savings through higher energy efficiency in the developing countries are at present somewhat similar to those in the industrialized countries.[9]

On the other hand, despite the rapid increase of commercial energy use in the developing countries, in the short to the medium run, the predominance of biomass fuels will remain. There are estimates to show that the scope for conservation of biomass is enormous in these countries.[10] In Bangladesh, for example, it is estimated that improved *chulahs* (stoves) can cut down wood fuel consumption by 50 to 80% with a modal value of 50%.

Means of energy efficiency gains

There are four ways that may be used to gain energy efficiency: housekeeping measures; fundamental process changes; product changes; and new energy conversion

---

7 Imran and Barnes, 1990.

8 World Bank, 1993.

9 Ewing, 1985; Fog and Nadkarni, 1983; Meunier and Kops, 1984; US Congress, 1992.

10 US Congress, 1992.

technologies.[11] Obviously, housekeeping measures may need little investment or organizational changes and may therefore be carried out rather quickly (see next section for details). Experience in countries like Japan shows that recovery of waste energy is an important step in increasing energy efficiency in many major industries like steel, petroleum and chemicals, cement, paper and pulp, dyeing and glass.[12]

4 Energy Conservation and Efficiency in Practice

What has been the actual experiences of countries in lowering their energy intensities? A brief discussion of the experiences will be presented first, after which some obstacles will be identified.

During the period 1970-1988 there was some improvement in world-wide energy intensity at the annual rate of 0.55%.[13] But not all groups of countries contributed positively, nor within each group did countries perform equally. Also, even within the same country not all sectors made a gain in terms of energy conservation.

Japan had an estimated energy intensity of 0.403 Kgoe/$ of real GDP in 1973. By 1988, it fell to 0.262 Kgoe/$, *i.e.* by more than 35%.[14] During the same period, the USA, West Germany and France lowered their energy intensities by 27%, 22% and 17% respectively. Among these countries the annual rates of improvement in energy intensity varied between 1.23% (Western Europe) and 1.83% (USA) to 2.22% (Japan).[15] During the same period, practically all other groups of countries showed little or no improvement or even large increases in energy intensity.

On the other hand, not all sectors of the economy lowered energy intensity equally. In most cases, the changes have been most apparent in industry, but there also have been other cases of major improvements. In Japan, for example, while electric refrigerators have become nearly 67% more efficient over 1973-1987, consuming only 26kWh/month in 1987 compared to 79.6 kWh/month in 1973, the improvement in the case of motor cars was nearly 50% (from around 9.4 Km/litre to about 13 Km/litre).

---

[11] Goldemberg et al., 1988.

[12] ESCAP, 1991, Table 5.1.

[13] ESCAP, 1991.

[14] Ogawa, 1992.

[15] ESCAP, 1991.

Even the same industry has not been able to exhibit equal efficiency among the industrialized countries. Steel is an example. While in 1987 the USA, France, Germany and the UK used very similar levels of energy per ton of crude steel produced (about 21 GJ/ton), they began from different bases in 1980 (ranging from 22.5 to 26 GJ/ton) and thus exhibited different rates of improvement. However, the energy consumption in production of crude steel in Japan had been much lower throughout this period. It even fell from 19.5 GJ/ton to 17.5 GJ/ton. These figures show that even among countries where information should flow most freely, there are obstacles to the adoption of the most efficient technology.

Over the 1980s, other developing or newly industrialized countries had also managed to lower energy intensity, particularly in industry. In China, where the overall energy intensity had risen over the 1970s, it fell by 31% over 1980-1988,[16] mainly because of a lowering of industrial energy intensity by similar percentages.[17] In Taiwan's manufacturing industries the intensity rose during 1970-1975, then fell, to an energy intensity level of just above 60% that of 1970. Practically all of this was due to pure intensity effect.[18] The Republic of Korea managed even better, with a fall of 44% in industrial energy intensity.[19]

Despite the above bright spots, many other developing countries could not improve their energy efficiency, and some even regressed. For example, while Thailand showed little or no change over 1970-1988, in Pakistan energy intensity increased by 26%, in India by 24% and in Malaysia by as much as 48%.[20] A comparable situation existed in countries of Latin America where, while some showed a decrease in energy intensity, most others did not and some like Mexico actually increased energy intensity during 1970-1980 over the preceding decade.[21] In Brazil, while some industries like cement improved energy use efficiency since 1973, others did not and the aggregate energy intensity increased by 20% over 1970-1988.[22]

Previously, the need for taking both direct and indirect requirements of energy into account, or of examining the issue on a system-wide basis when making sectoral

---

[16] Imran and Barnes, 1990.

[17] Huang, 1993.

[18] Li, Shrestha and Foell, 1990.

[19] Park, 1992.

[20] Imran and Barnes, 1990.

[21] Guzman et al., 1987.

[22] Imran and Barnes, 1990.

comparisons, has been emphasized. With the exception of rather few studies, this issue has hardly been given attention in the literature. A study on India, which looked into energy use as a part of the analysis of CO_2 emission found that if direct CO_2 emissions from fossil fuel use are considered, then electricity generation tops the list, being responsible for a third of the total emissions.[23] Iron and steel ranks second and other transport (excluding railways) ranks third, contributing only 9.8% and 8.6% respectively. Per 1,000 residences the CO_2 emission of electricity generation is 3.47 tons. Considering indirect usage of fuels, the coefficient goes up to 4.89, *i.e.* by more than 40%. In case of construction activities - which utilize cement, iron and steel, *etc.* on a large scale - the direct coefficient is only 0.003 tons, increasing to 0.862 tons when both direct and indirect emissions are taken into consideration. As a result the construction sector emerges as the largest CO_2 emitting sector in India, accounting for 22% of the total followed by food crops production (8%) and other transport (nearly 6%). Of course CO_2 emission per unit of output does not directly tell us someting about energy intensity. However, because of the intimate link between the two, these results can be used as a pointer in the relative ranking of sectors in terms of aggregate energy efficiency. Another study on Australia did not compare the direct and total (direct plus indirect) coefficients but clearly shows that not all sectors which appear to be CO_2 emission-efficient (carbon per unit of final demand) are low emitters in aggregate. Again the construction sector is a very low emitter on emission-efficiency basis, but a major one on a total emission basis. As these emission figures depend very much on energy efficiencies and use, similar conclusions are likely to emerge if inter-industry demands are included in sectoral comparisons. In any case what these results signify is that energy efficiency is not enough in saving energy. One will also have to give attention to development of technologies which use less energy-intensive products as inputs in other processes like construction activities.

Role of policies and other factors

The experiences as briefly discussed above show a mixed picture. While some of the developing countries had certainly been less successful or even failed to increase energy efficiency, even in the industrialized countries the constraints to achieving higher energy efficiency were many. To understand these factors let us first clarify the issue of cost-effectiveness. Recent research has amply shown that there are many

[23] Parikh and Gokarn, 1993.

energy-saving ideas and techniques which are available at little or no extra cost.[24] Some of these are really very simple measures like painting roofs white and planting shade trees around houses so that demand for energy (such as for air-conditioning) falls.

Several measures have been taken in various countries for encouraging energy conservation. These measures, while good on paper, were not always easy to implement and the intended effects have not always been achieved or clearly discernible.

The energy commodity whose price sets the trend for those of others is oil. Currently, the price of oil is still depressed but higher than the long-run marginal cost of its extraction. This of course leads to complacency and a lack of political will in major consuming nations to raise prices and to consequent consumer resistance. One way in which the prices of oil can be raised is to tax imported oil. This, however, creates dissatisfaction among the OPEC countries (note the recent vehement reaction by OPEC countries to the proposal of a carbon tax). Despite such problems countries like Sweden, have progressed far more than many others in moving away from cheap energy, while countries like the USA have lagged far behind in terms of political will and business, and consumers' resistance.[25]

When it comes to the issue of adoption of more efficient energy using devices and equipment, the experience is not always so clear cut. The reasons are varied. While on the one hand the hypothesized adverse impact of rising energy prices is not readily discernible,[26] environmental regulations have been found to have resulted in a lower productivity in some industries in the USA.[27] On the other hand, others also have found that once prices of energy have risen and firms adjusted to the price rise, a subsequent decline in price and then a rise from the lower price level would not affect output adversely.[28] If true, upward influence on price of energy through fiscal means is not likely to be politically difficult to achieve in many industrialized countries where adjustments have already been made before the present low level of oil prices. Results of simulations in developing countries like the Philippines indicate that the manufacturing sector may have adjusted substantially to the taxes imposed in the wake of the first oil shock, and therefore the elimination of these taxes will not

[24] Rubin et al., 1992; Flavin and Durning, 1988; US Congress, Office of Technology Assessment, 1992.

[25] Bina, 1989; Landsberg, 1988.

[26] Hisnanick and Kymn, 1992.

[27] Barbara and McConnel, 1990.

[28] Smyth, 1993.

result in appreciable change in manufacturing output, although households may increase energy consumption substantially.[29]

To better place the issue in perspective, a point or two should here be raised about the observed price and income elasticities for energy demand. There are controversies over these estimates,[30] but the general conclusion is that the short-run elasticities in the industrialized countries are low. A pooled estimate for USA, Germany, Japan, France and UK showed a short-run price elasticity of only -0.13 and an income elasticity of 0.24.[31] The respective long-run elasticities are -0.34 and 0.63. Similar results have been obtained by others with much higher income elasticities, sometimes close to or beyond unity.[32] The demand for energy therefore may not be as sensitive to price rises as one would like to assume, although estimates for the USA show that at times this could be close or equal to *minus* unity.[33]

In the developing countries, price elasticities have been found to be low both in the short and the long run with median values around -0.15 and -0.25. Income elasticities on the other hand are mostly above or close to unity and sometimes very high (1.63) for specific energy consumption like oil in Morocco.[34] In developing countries therefore, while in general price elasticities are higher than or at most equal - in absolute value - to those observed in the industrialized countries, the income elasticities are far higher, probably indicating a high level of unsatisfied demand.

A question may arise here as to whether energy price or the price of the equipment through which energy is used is more important. A study has shown that in Germany it is the latter which is more important in household decision making to invest in more energy efficient devices and thus to save energy.[35] The effort of the households in using less energy can only be successful when more energy efficient equipment is known to be available and affordable. What all these results indicate is that reliance on the market through energy prices is necessary but inadequate to influence energy consumption fully in the desired direction without supplementary policies, both in the industrialized and possibly even more so in the developing countries.

---

29 Uri and Boyd, 1993.

30 Hsing, 1990, and references therein.

31 Welsch, 1989.

32 Bentzen and Engsted, 1993; Koshal et al., 1990.

33 Hsing, 1990.

34 Ibrahim and Hurst, 1990.

35 Flaig, 1990.

Another point which merits attention here is the debate over whether capital and energy are complements or substitutes.[36] Complementarity, on the one hand, means that if energy prices rise and energy consumption falls, the demand for capital services will also fall and thus lower output. A substitution, on the other hand, increases the demand for capital and need not lower output. In general time series analyses show the relationship to be complementary while pooled cross-section studies show these to be substitutes. This has been interpreted to mean that energy and capital are complements in the short run and substitutes in the long run.[37] If accepted, the energy-capital controversy does not pose any problem in terms of energy use and conservation measures. As energy use efficiencies are generally embodied in the machines and equipment, in the short run energy and capital must be complements. In the longer run, as newer machines and processes with higher embodied energy efficiency come into the market and are used in production, the technological change exhibits substitutability. In any case the upshot is that the issue is important and should be clearly understood in each country's own context with as much disaggregation as possible on an industry-by-industry basis.

Apart from prices, there had been measures like credit programmes and various regulatory practices. In the household sector, while credit programmes have largely failed in one country,[38] they had better success elsewhere.[39] The latter study found that expenditure on conservation measures are positively correlated with income and availability of federal credit. In the case of demand for automobile efficiency, it has been found that demand for fuel efficiency is highly elastic. But while an increase in price of gasoline would increase the demand for more efficient vehicles, the demand for other vehicle attributes like general comfort of ride and of front seat comfort also rise.[40] Satisfying demand for both energy efficiency and such other attributes may take longer than simply satisfying energy efficiency, thus slowing down the rate of growth of the latter.

It appears that despite certain improvements, the rise in energy efficiency has been lower than its full potential, even under the present technological know-how. Consumer (households, firms) behaviour suggests a very high rate of discounting (*i.e.* a rather short pay-back period). At these discounting rates, the potential for energy

[36] Apostolakis, 1990.

[37] Apostolakis, 1990.

[38] Lafrance and Perron, 1993.

[39] Long, 1993.

[40] Atkinson and Halversen, 1984.

savings diminishes substantially. Some attempts have been made to explain the situation[41] and there are several implications of these analyses. Some of these and other relevant issues are discussed in the next section.

5 Future Policy Directions

With respect to future energy policy it has often been suggested that a suitably designed tariff structure can improve rationalising energy demand and increase equity in both industrialized and developing countries.[42] However, from the literature it can be concluded that a good energy policy should involve more than simply getting prices right. Indeed, the failure to realize the full potential of energy saving has been considered as a market failure. Information about the costs, reliability and performance of efficient technologies are not readily available, calling for monitoring and certification about performance of energy-using equipment and its proper labelling, and for a public education and awareness campaign. There is some evidence that these do have good pay-offs and that the discount rate used by, say, households while purchasing properly labelled and advertised energy-efficient houses comes down to quite plausible levels of approximately 8%.[43] It should be indicated that prices of energy-using equipment may be more influential than energy prices in decision making. If the required investments are substantial, then suitable credit programmes may help in such cases.

Taxes and subsidies, if suitably designed, can help in achieving a desirable level of energy efficiency. These instruments can eventually be combined with other measures; for example, the money saved or generated by withdrawal of subsidies on or taxing energy consumption may be used to finance credit/subsidy programmes for the manufacturing/purchasing of energy-efficient equipment. Before any such measure is accepted and made operational, the pros and cons need, of course, to be clearly analyzed.

For example, the introduction of a carbon tax has been politically resisted (particularly in the USA and by the OPEC), since its effects appear uncertain and regressive in both industrialized and developing countries.[44] Furthermore, while imposing taxes

[41] Howarth and Anderson, 1993; Rubin et al., 1992.

[42] Brand, 1990; Markandya and Pemberton, 1990.

[43] Horrowitz and Haeri, 1990.

[44] Agostini et al., 1992; Shah and Larsen, 1994.

one needs to be sure about the nominal and effective rates. In Bangladesh, for example, the nominal tax rates are 4%, 0% and 67% on, respectively, petroleum products, electricity, and natural gas. The estimated effective rates when inter-industry flows and capital taxation are considered, however, rise to 42%, 32% and 72%, respectively.[45] Further taxation on these energy sources can probably not be called for. Also, while taxes and subsidies may be used to bring 'proper equilibrium' in the market and may have impacts in the intended direction without major dislocations elsewhere,[46] the information requirements for these to be effective are in general onerous.

But more importantly, the deck needs to be cleared before new measures are announced. For example, the removal of substantial fossil fuel subsidies (existing in many countries) is sometimes estimated to be equivalent to a carbon tax of $40 per ton of CO_2[47] in terms of generation of funds.

The limitations of the market mechanism with respect to energy efficiency suggest an important role for direct regulation of equipment performance and energy-efficiency standards; any charge levied on the landlord or the property is otherwise likely to be shifted on to the tenant. The cost of monitoring and enforcing of standards in such cases, however, may be quite high. Tradeable permits in energy-efficiency in such circumstances have been put forward as an efficient method of enforcement. The general idea is quite appealing. The merit of such proposals or how such a policy should be designed is, however, still being debated.[48]

In developing countries, raising energy efficiency faces problems similar to those described above. In some cases, the response underlying observed behaviour in times of energy price changes sometimes appears rather muted,[49] while the rates of an effective tax in some cases may already be quite high. These are all the more reason for direct regulatory measures in such situations.

How to increase energy efficiency in the consumption of wood fuels? Here institutional measures and proper extension are probably quite important. Popularising energy-efficient cooking stoves among hundreds of thousands of households would

---

45 Chowdhury, 1994.

46 Fisher and Despotakis, 1989.

47 Shah and Larsen, 1994.

48 Malueg, 1990; Malik, 1990; Keeler, 1991; Malik, 1993.

49 Mahmud and Chisti, 1990; Siddayao et al., 1987.

necessitate efforts at many levels. But here too, suitably designed credit and, if necessary, subsidies or tax breaks may help in manufacturing the new stoves in large numbers.[50] However, dissemination may be difficult. Non-government efforts in this area may go a long way in solving the problem.

Finally, some remarks concerning the possible impacts of some of the measures may be noted. Environmental regulations in a large trading country or a number of trading countries can have major domestic and international repercussions.[51] While international negotiations to solve the problems may be initiated, these will take long periods to be completed, whereas probably not all countries may wish to participate. The suggested solutions in such cases become increasingly more complex[52] and possibly politically unpalatable. Whether mechanisms like JI can help in such situations still remains to be worked out, but it seems that, with respect to developing countries, there are areas where JI can play a role.

One reason for the slow improvement in energy efficiency is the lack of local skills for adaptation of technology. A part of the deferred expenditure for investment in energy efficiency, which cannot be carried out for some reason in the industrialized countries, could be transferred to the developing countries and jointly managed to develop local skills and to install more efficient equipment in local factories. This is likely to be far more cost-effective than equivalent improvement in energy efficiency in the industrialized nations. A major condition that such transfers should fulfil to be acceptable to the developing nations is that these should be wholly additional to the traditional flow of aid and other similar resources. Indeed, these transfers need not be part of government-to-government resource flows but may be between enterprises in the respective countries and should therefore not count as part of the traditional resource aid.

---

[50] Hurst, 1990.

[51] Krutilla, 1991.

[52] Bohm, 1993.

Annex. Absolute vs. relative use of energy.

Let

E = energy,
GDP = gross domestic product,
P = population
e = average energy intensity of GDP (= E/GDP)
y = per capita GDP (= GDP/P).

Then,

$$E = e \times y \times P \tag{1}$$

Taking logarithms and total differentials on both sides, we get

$$(1/E)\, dE = (1/e)\, de + (1/y)\, dy + (1/P)\, dP \tag{2}$$

which reduces to

$$dE = (E/e)\, de + (E/y)\, dy + (E/P)\, dP \tag{3}$$

Equation (3) shows that the change in total energy use depends on three factors, viz. change in energy intensity of GDP, the change in per capita GDP, and the change in population. Equation (2) shows the same relationship in terms of rates of change. Equation (3) also shows how the absolute and the two related measures of energy use (relative to income and population) are interrelated.

Quite obviously, total energy use may fall if energy use per unit of GDP, or per capita GDP or population, falls (in each case, all other things remain unchanged). As a fall in GDP per capita is an inconceivable policy option and population levels may fall only in extreme circumstances, the major policy option that remains is a fall in energy intensity of GDP or an increase in energy efficiency of economic activities. In terms of the rate of change equation, the lower the rate of growth of per capita GDP and population and the higher the improvement in energy intensity, the lower will be the rate of growth of total energy use.

References

Abdel-Khalek, G., 1988. 'Income and Price Elasticities of Energy Consumption in Egypt : A Time Series Analysis', *Energy Economics*, January 1988.

Agostini, P., M. Botteon, and C. Carraro, 1992. 'A Carbon Tax to Reduce CO2 Emissions in Europe', *Energy Economics*, October 1992.

Apostolakis, B.E., 1990. 'Energy-capital substitutability/complementarity: the dichotomy', *Energy Economics*, January 1990.

Atkinson, S.E. and R. Halversen, 1984. 'A New Hedonic Technique for Estimating Attribute Demand: An Application to the Demand for Automobile Efficiency', *The Review of Economics and Statistics*, Vol. LXVI, No. 3, August 1984.

Barbara, A.J. and V.D. McConnel, 1990. 'The Impact of Environmental Regulations on Industry Productivity: Direct and Indirect Effects', *Journal of Environmental Economics and Management*, Vol. 18, No. 1, January 1990.

Barnett, A., 1990. 'The Diffusion of Energy Technology in the Rural Areas of Developing Countries: A Synthesis of Recent Experience', *World Development*, Vol. 18, No. 4, 1990.

Belgrave, R., Ch.K. Ebinger and H. Okino (ed.), 1987. *Energy Security to 2000*, Gower, Aldershot and Westview Press, Boulder.

Bentzen, J. and T. Engsted, 1993. 'Short- and long-run elasticities in energy demand: A cointegration approach', *Energy Economics*, January 1993.

Bina, C., 1989. 'Competition, controls and price formulation in the international energy industry', *Energy Economics*, July 1989.

Bohm, P., 1993. 'Incomplete International Cooperation to Reduce CO_2 Emissions: Alternative Policies' *Journal of Environmental Economics and Management*, Vol. 24, No. 3, May 1993.

Brand, M., 1990. 'Impacts of the structures of electricity and gas tariffs on the rational use of energy', *Energy Economics*, April 1990.

Chandler, W.W., 1985. *Energy productivity: Key to Environmental protection and Economic Progress*, Worldwatch paper 63, 1985.

Chowdhury, O.H., 1994. *Incidence of Indirect Taxation in Bangladesh : 1984/85*, Research Monograph 16, BIDS, Dhaka.

Common, M. S. and U. Salma, 1992. 'Accounting for changes in Australian carbon dioxide emissions', *Energy Economics*, July 1992.

Eden, R., 1991. 'Energy Conservation : Opportunities, Limitations and Policies' in: Mabro (ed.), 1991. ESCAP (Economic and Social Commission for Asia and the Pacific), *Energy Policy Implications of the Climatic Effects of Fossil Fuel Use in the Asia-Pacific Region*, Bangkok.

Ewing, A.J., 1985. *Energy Efficiency in the Pulp and Paper industry with Emphasis on Developing Countries*, World Bank Technical Paper No. 34, Washington D.C., 1985.

Fisher, A.C. and K.A. Despotakis, 1989. 'Energy Taxes and Economic Performance : A Regional General Equilibrium Analysis', *Energy Economics*, April 1989.

Flaig, G., 1990. 'Household production and the short and long run demand for electricity', *Energy Economics*, April 1990.

Flavin, C., 1988. 'Creating a Sustainable Energy Future', *State of the World, 1988*, A Worldwatch Institute Report on Progress Toward a Sustainable Society, W. W. Norton & Company, New York and London.

Flavin, C. and A. Durning, 1988. 'Raising Energy Efficiency', *State of the World, 1988*, A Worldwatch Institute Report on Progress Toward a Sustainable Society, W. W. Norton & Company, New York and London.

Fog, M.H. and K.L. Nadkarni, 1983. *Energy Efficiency and Fuel Substitution in the Cement Industry with Emphasis on Developing Countries*, A World Bank Technical Paper, Washington D.C.

Gilmer, R.W., 1989. 'Energy Labels and Economic Search: An Example from the Residential Real Estate Market', *Energy Economics*, July 1989.

Goldemberg, J., et al., 1988. *Energy for a Sustainable World*, Wiley Eastern Limited, New Delhi.

Guzman, O., A. Yunez-Naude and M.S. Wionczek, 1987. *Energy Efficiency and Conservation in Mexico*, Westview Press, Boulder and London.

Hisnanick, J.J. and K.O. Kymn, 1992. 'The impact of Disaggregated Energy on Productivity : A Study of the US Manufacturing Sector, 1958-85', *Energy Economics*, October 1992.

Horowitz, M. J. and H. Haeri, 1990. 'Economic efficiency v energy efficiency: Do model conservation standards make good sense', *Energy Economics*, April 1990.

Howarth, R.B. and B. Anderson, 1993. 'Market Barriers to Energy Efficiency', *Energy Economics*, October 1993.

Hurst, C., 1990. 'Establishing New Markets for Mature Energy Equipment Developing Countries: Experience with Windmills, Hydro-powered Mills and Solar Water Heaters', *World Development*, Vol. 18, No. 4.

Hsing, Y., 1990. 'On the variable elasticity of demand for gasoline', *Energy Economics*, April 1990.

Huang, J-P., 1993. 'Industry Energy Use and Structural Change: A Case study of the People's Republic of China', *Energy Economics*, April 1993.

Ibrahim, I. B. and C. Hurst, 1990. 'Estimating energy and oil demand functions: A study of thirteen developing countries', *Energy Economics*, April 1990.

Keeler, A.G., 1991. 'Non-compliant Firms in Transferable Discharge permit markets: Some Extensions', *Journal of Environmental Economics and Management*, Vol. 21, No. 2, September 1991.

Koshal, R. K. et al, 1990. 'Production and high energy prices: A case of some Pan-Pacific countries', *Energy Economics*, July 1990.

Kosmo, M., 1987. *Money to Burn? The High Costs of Energy Subsidies*, World Resources Institute.

Krutilla, K., 1991. 'Environmental Regulation in an Open Economy', *Journal of Environmental Economics and Management*, Vol. 20, No. 2, March 1991.

Kumar, M.S., 1987. *Energy Pricing policies in Developing Countries: Theory and Empirical Evidence*, UNDP, ESCAP and ILO-ARTEP, ILO, Geneva.

Lafrance, G and D. Perron, 1993. 'Energy Demand Impact of Incentive Programmes: A Non-parametric Analysis for the Canadian Oil Substitution Programme', *Energy Economics*, October 1993.

Landsberg, H.H., 1988. *Energy Policy in Historical Perspective*, Working paper 88-13, Colorado School of Mines, Department of Economics, Golden, Colorado.

Li, J., R. M. Shreshtha and W. K. Foell, 1990. 'Structural change and energy use: The case of the manufacturing sector in Taiwan', *Energy Economics*, April 1990.

Long, J.E., 1993. 'An Econometric Analysis of Residntial Expenditures on Energy Conservation and Renewable Energy Sources', *Energy Economics*, October 1993.

Mabro, R. (ed.), 1980. *World Energy: Issues and Policies*, Oxford University Press, 1980.

Mahmud, F. and S. Chisti, 1987. 'The Demand for Energy in the large-scale Manufacturing Sector of Pakistan', *Energy Economics*, April 1987.

Malik, A.S., 1990. 'Markets for Pollution Control When Firms are Non-compliant', *Journal of Environmental Economics and Management*, Vol. 18, No. 2, March 1990.

Malueg, D.A., 1990. 'Welfare Consequences of Emission Credit Trading programs', *Journal of Environmental Economics and Management*, Vol. 18, No. 1, January 1990.

Markandya, A. and M. Pemberton, 1990. 'Non-linear prices and energy demand', *Energy Economics*, January 1990.

Meunier, M.Y. and Oscar de B. Kops, 1984. *Energy Efficiency in the Steel Industry with Emphasis on Developing Countries*, World Bank Technical Paper No. 22, Washington, DC.

Ogawa, Y., 1992. 'Analysis of Factors Affecting Carbon Dioxide Emissions due to Past Energy Consumption around the World', ESCAP, *Green House Research Initiatives in the ESCAP Region: Energy*, Bangkok.

Parikh, J. and S. Gokarn, 1993. 'Climate change and India's energy policy options: New perspectives on sectoral CO_2 emissions and incremental costs', *Global Environmental Change*, September 1993.

Park, S.-H., 1992. 'Decomposition of Industrial Energy Consumption: An Alternative Method', *Energy Economics*, Vol 14, No. 4, October 1992.

Pearson, M. and S. Smith, 1991. *The European Carbon Tax: An Assessment of the European Commission's Proposals*, The Institute of Fiscal Studies, London.

Rogers, E.M. and F.F. Shoemaker, 1971. *Communications of Innovations: A Cross Cultural Approach*, Free press of Glencoe, New York.

Rubin, E.S. et al, 1992. 'Realistic Mitigation Options for Global Warming', *Science*, Vol. 257, July 10.

Shah, A. and B. Larsen, 1994. *Global Warming, Carbon Taxes and Developing Countries*, a paper read at the IPCC working Group III workshop on policy instruments and their implications, Tsukuba, 17-20 January 1994.

Siddayao, C. M., M. Khaled, J. G. Ranada and S. Saicheua, 1987. 'Estimates of Energy and Non-energy Elasticities in Selected Asian manufacturing Sectors: Policy Implication', *Energy Economics*, April 1987.

Smyth, D.J., 1993. Energy Prices and the Aggregate Production Function', *Energy Economics*, April 1993.

US Congress, 1992. Office of Technology Assessment, *Fuelling Development: Energy Technologies in Developing Countries*, OTA-E-516, US Government Printing Office, Washington D.C., April 1992.

Uri, N.D. and R. Boyd, 1993. 'An Evaluation of the Effects of the tax on Refined Petroleum Products in the Philippines', *Energy Economics*, July 1993.

Walsh, M.J., 1989. 'Energy Tax Credits and Housing Improvement', *Energy Economics*, October 1989.

Welsch, H., 1989. 'The Reliability of Aggregate Energy Demand Functions: An Application of Statistical Specification Error Tests', *Energy Economics*, October 1989.

World Bank, 1993. *Energy Efficiency and Conservation in the Developing World*, A World Bank Policy Paper, Washington, DC.

20. Energy Efficiencies of Industrial Processes and Electricity Production in European and Non-European Countries

KORNELIS BLOK, DIAN PHYLIPSEN, ANDRE FAAIJ and ERNST WORRELL*

1 Introduction

It is well-known that large differences in energy consumption and energy efficiency between countries exist. Up to now comparisons between countries have been made on a national level, based on aggregates, such as energy consumption per capita or per unit of GDP.[1] Although such comparisons are adequate to obtain a first impression of the order of magnitude of the differences in energy efficiency between countries, they do not give much insight in the causes of the international differences, nor in the ways to achieve a lower energy consumption level.

Here, more detailed comparisons of energy efficiency between countries will be made. First, an overview will be given of the energy consumption by main sector for various groups of countries. Subsequently, we will discuss how comparisons of energy consumption between countries for various main sectors can be made. Finally, we will evaluate these comparison results for policy formulation.

2 Breakdown of Energy Consumption by Main Sectors

What sectors are the most important ones with respect to energy consumption? In Figure 20.1 an overview is given of the energy requirements for various countries or groups of countries. The first bar for each country represents the total primary energy consumption (including the electricity production) per sector.[2] The second bar represents the distribution of the electricity use by sector. The transformation and

* The authors are associated with the Department of Science, Technology and Society, Utrecht University, The Netherlands. They wish to thank the National Research Programme on Global Air Pollution and Climate Change (NOP-MLK) for the financial support of the research presented in this chapter.

[1] OECD, 1991; Nilsson, 1993.

[2] For reasons of presentation the category 'other sectors' has been omitted.

C. J. Jepma (ed.), The Feasibility of Joint Implementation, 285–297.

distribution losses occurring in electricity production are proportionally divided over the electricity consuming sectors.[3 4 5]

From Figure 20.1 it can be inferred that the level of energy consumption and the breakdown into main sectors strongly differ between various countries.

Figure 20.1. Comparison of energy consumption per capita by sector in several countries or groups of countries for 1990.[a]

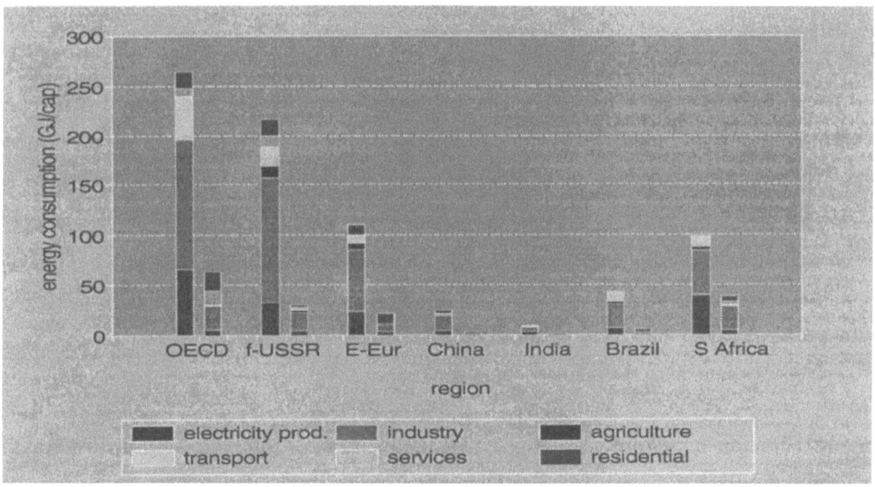

[a] The figures for industry include oil refineries, gas works, coal and other transformation. Electricity production includes auto producers of electricity and CHP plants. District heating has been included in the residential sector.

3 International Energy Efficiency Comparisons

The level of energy consumption in an economic sector is determined by:

- the total level of human activity within the sector;

---

[3] The sectors mentioned in Figure 20.1 are those used in the OECD statistics and energy balances; IEA, 1993a; IEA, 1993b.

[4] The figures for the Eastern European region have been calculated by adding up the figures for the individual countries (Albania, Bulgaria, former Czechoslovakia, Hungary, Poland, Romania and former Yugoslavia).

[5] IEA, 1993b.

- the mix of activities within the sector (the structure); and
- the energy efficiency within the sector (the energy consumption per unit of activity).[6]

All of these factors can be the subject of a policy to reduce energy-related GHG emissions; improving energy efficiency can be considered to be the most important of them in the short term.

With respect to energy efficiency one can distinguish between the efficiency of energy conversion activities and of energy end-use activities. In general, the efficiency measure of energy conversion used is energy output per unit of energy input. For energy end-use, however, the situation is more complicated: the level of human activity can be measured either in monetary units (*e.g.* value added of the steel industry) or in physical units (*e.g.* tons of steel produced). This leads to two measures of (actually the inverse of) energy efficiency:

1. energy intensity, *i.e.* energy consumption per unit of value added;
2. specific energy consumption, *i.e.* energy consumption per physical unit of human activity (*e.g.* m^2 of space heated, person-Km of transportation, tons of steel produced).

The second measure generally provides a better insight in the technological characteristics of the use of energy, because the first measure is also affected by other factors, such as feedstock and product prices. However, the second measure is not applicable to all sectors (like the light industries) because not for all sectors a good physical indicator of human activity can be defined. In this analysis, which focuses on heavy industrial sectors the second indicator will be used.

The next subsections will present some energy efficiency results for: a number of heavy industrial sectors; the electricity production sectors; and the transportation and residential sector. The analysis of the latter two sectors will be based on the results from other studies.

3.1 Energy efficiency in heavy industry

In the 'heavy industry' sectors that produce intermediate products like metals, basic chemicals, pulp and paper, and building materials, the activity level can generally be

---

[6] Schipper, Meyers, Howarth and Steiner, 1992.

measured in tons of products. Hence, the obvious energy efficiency measure is the energy consumption per unit of product.

From an analysis comparing energy consumption by industrial processes in countries of the EU,[7] it has been concluded that the mix of feedstock is an important structural factor explaining energy consumption, especially the extent to which primary or secondary materials are used. Another structural factor that has been identified is the product mix. These factors are therefore being taken into account when comparing the present Specific Energy Consumption (SEC) of a country with a 'best practice' SEC in order to determine the potential for energy efficiency improvement; 'best practice' SEC is defined here as the lowest SEC observed in a sector or plant in the EU in the reference year (*i.e.* 1988). It should be noted that the SEC as well as the 'best practice' SEC are time dependent, so that the total future potential for energy efficiency improvement may differ from the values mentioned in this chapter.

On the basis of additional information,[8] the analysis can be extended to countries outside the EU. The results for ammonia production, cement production and steel production will be presented in Annex 1 and are depicted in Figures 20.3, 20.4 and 20.5 respectively. Both the 'best practice' SEC and the present SEC are given; the difference between the two represents the energy efficiency improvement potential.

3.2 Energy efficiency in fossil fuel-based electricity production

With respect to the electricity production sector, the energy efficiency is defined as the energy content of the electricity produced divided by the energy input per type of fossil primary energy carrier. The energy efficiency of electricity produced by all types of fossil primary energy carriers is expressed as a weighed average of these efficiencies on the basis of produced electricity. The study on the energy efficiency potential in electricity production has been limited to the public electricity production sector because of a data limitations. Distribution losses are not taken into account as we want to compare the efficiencies of the power plants only.

---

[7] Worrell, Cuelenaere, Blok and Turkenburg, 1995.

[8] Bossenbroek, 1993; Garcia del Valle, 1990; Meyers et al., 1993; Ross et al., 1991; Statistical information on Czech Republic, Slovak Republic and Poland, 1990; International Iron and Steel Institute, 1990 and 1992; Vallance, 1993.

Figure 20.2. Efficiency of central fossil based electricity production in the 12 countries with the highest amount of fossil-fuel based electricity production.[a]

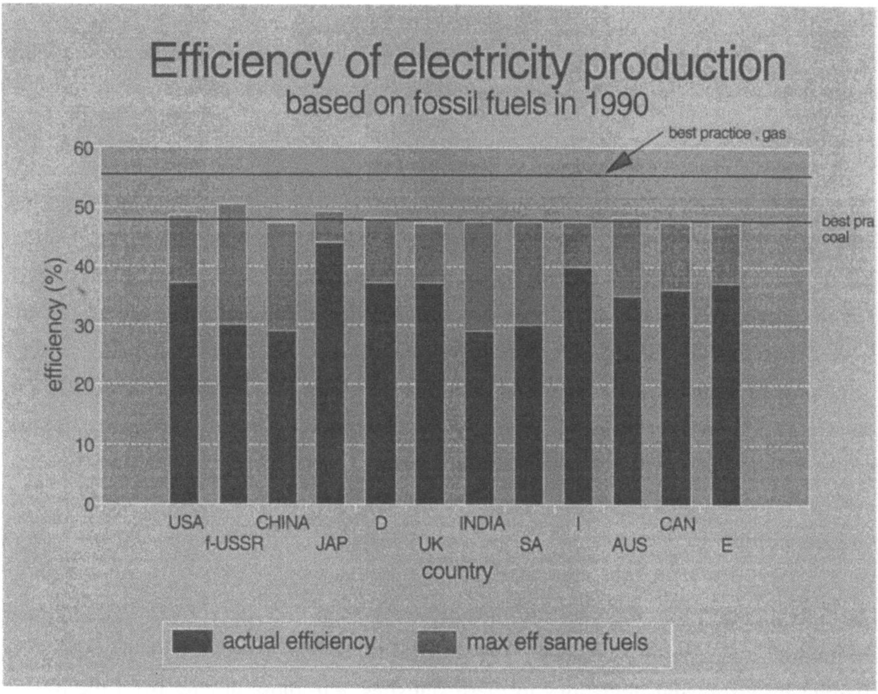

[a] The efficiencies are corrected for effects of CHP on electricity generating efficiency. The bars represent the maximum efficiency achievable with the present fuel mix in the given regions and countries. The dark parts of the bars depict the present efficiency. The two solid lines represent the maximum efficiency at present for gas-fired and coal-fired electricity plants.

The figures presented here are only valid for public electricity generation; auto producers are excluded from the analysis. Part of the public electricity production is combined with the production of heat (combined heat and power generation, or CHP); in general, the heat is distributed via district heating systems. A power plant that operates in the CHP mode generally has a somewhat lower electric efficiency, due to the extraction of heat. In order to correct for this effect, we assume that each unit of heat that is extracted causes a loss of 0.175 units of electricity production (both on an enthalpy basis).[9] It should be noted that in this way a fair base for a comparison of the quality of the energy conversion of the power plants is obtained.

[9] For further details see Faaij et al., *forthcoming*.

No credit is given to the fact that the application of CHP as such is already a way of saving on primary energy. This would require a separate analysis.

The results of the analysis are depicted in Figure 20.2. The countries selected in this figure are the largest producers of fossil-fuel based electricity and are presented in descending order of fossil-fuel based electricity production. Due to large uncertainties in the data used in this study results should be interpreted with care.

3.3 Energy efficiency comparisons in other sectors

Other sectors studied with regard to energy efficiency are transportation and residential. In the studies energy intensity is used as an indicator of energy efficiency, even if it is determined by the mix of the level of activity, the structure of the sector and the (technological) energy efficiency of the sector. The (technological) energy efficiency of transportation is expressed in person-Km (for passengers) or ton-Km (for freight). For the residential sector ways of expressing energy efficiency depend on the function studied (*e.g.* space heating, cooking, *etc.*).[10]

3.4 Discussion

The methodology in this section heavily relies on the data from international production and energy statistics. The reliability of the results will therefore be affected by errors and deviations in these statistics, *e.g.* those arising from the use of different methodologies in national statistics, or from the non-uniform definition of boundaries of sectors. Moreover, stocks of primary energy carriers and electricity produced with CHP, for instance, are sometimes treated differently; figures are occasionally reported less frequently than once a year, so that intermediate data are to be obtained by extrapolation. Often only aggregated data are available or are not available at all; some countries' statistics are unreliable.

The IEA indicates that although the figures on total fuel and electricity consumption for most OECD countries are regarded accurate, they still have difficulties in giving a detailed breakdown by fuel for industry consumption. This is especially true for the transformation sector (refineries, electricity production). Comparing several statistical sources, Worrell et al. estimated the influence of statistical errors to be 12% of the

---

[10] For the main energy function in the residential sector, i.e. space heating, comparisons have been made between OECD countries, Eastern European countries and developing countries by Schipper et al., 1992; 1993; Howarth et al., 1991.

results for the oil refining sector.[11] For non-OECD countries the accuracy of the statistics is in general rather poor.

In order to improve the results of international analyses of energy efficiencies, it is therefore needed to improve the available statistics in an internationally harmonized way; there is a strong need for the design of common methodologies to calculate energy efficiencies.[12] The developed methodology also needs to be tested for applicability in other sectors, especially in relation to other energy intensive sectors than those discussed in this chapter.

4 Application of International Energy Efficiency Comparisons

The following applications for which the comparisons presented in Section 3 can form the basis can be distinguished.

Improvement of the knowledge on potentials and costs of fostering energy efficiency
The levels of energy conversion efficiency and specific energy consumption give a first indication of the reduction potential that is still available. However, it is not yet clear how much the costs will be for each of the countries to reach a certain level of efficiency. In order to obtain that information further studies have to be carried out, as this has already been done in various countries. In general, it is difficult to compare the outcomes of these studies. For instance, some comparisons of costs have been made at a national level,[13] but then it is not exactly clear whether the strong differences between the results are real or just caused by differences in starting points (*e.g.* differences in the set of technological options that is considered). International comparisons of energy efficiency, followed by a careful sector-by-sector comparison of costs therefore forms a more reliable basis for getting a better understanding of the real differences in potentials and costs for energy efficiency improvement.

---

[11] Worrell et al., 1995, *forthcoming*.

[12] At present some projects are underway, such as the project 'International comparisons of energy efficiency' that is coordinated by the Lawrence Berkeley Laboratory and Utrecht University (The Netherlands), in which a number of international experts in the field of energy efficiency cooperate. Another project is the 'Cross-country comparisons of energy efficiency indicators' coordinated by ADEME (France).

[13] Gusbin, 1991; Kram, 1993; Swisher et al., 1992.

International agreements on energy efficiency levels

Up to now not much progress has been made in reaching international agreements on emission reduction of CO_2. Several rather normative and general approaches, based on equal relative emission reduction or equal emissions per unit of GDP suggested, have led to objections from some of the countries involved in the negotiation processes. A better approach therefore seems that countries conclude agreements on energy efficiency at a sectoral rather than at the national level. For example, for an industrial sector (*e.g.*, the iron and steel industry) countries could agree on reaching the 1990 'best practice' level in the year 2000. This would require to determine the *average* level of the SEC for that sector in the countries involved in the agreement, whereby the structural differences (feedstock and products) between countries should be taken into account.

This way one could more or less follow the fruitful procedure followed in the negotiations on the reduction of the emission of substances that deplete the ozone layer: starting from a limited set of main substances the protocols have gradually been extended and strengthened.

International technological cooperation

International cooperation in the introduction of technologies that reduce the emissions of GHGs is one of the main issues in the climate negotiations, especially in the field of Joint Implementation (JI). Also here international comparisons of energy efficiency can make an important contribution as an initial step in steering the cooperation activities. These comparisons provide an indication of which sectors in which countries should have the highest priorities. Furthermore, they can indicate which type of transfer is needed most: investment capital, knowledge and education, licences, *etc.*

5 Conclusions

International comparisons of energy efficiency can be carried out for many sectors, covering a considerable part of world energy demand. Development of common methodologies of measuring energy efficiency and improving the availability and quality of data is necessary. Such a task seems worthwhile, the more so when taking into account the possible applications of international energy efficiency comparisons in an international climate policy framework.

Annex 1. Comparison of the Specific Energy Consumption (SEC) in ammonia, cement and steel production.

Ammonia production

Ammonia can be produced by partial oxidation of oil residues and by steam reforming of natural gas. The specific energy consumption of the first process typically amounts to 30 GJ/ton ammonia for a modern plant and up to 39 GJ/ton for older ones. Of these two techniques, steam reforming of natural gas is the most energy efficient process. Worldwide, 80% of the ammonia is produced by steam reforming of natural gas, while partial oxidation accounts for about 14% of the world's ammonia production capacity. In the EU only 8% of the ammonia production capacity is based on the partial oxidation process. The 'best practice' SEC of ammonia has been determined at 28 GJ/ton ammonia and has been derived from the ICI-AMV steam reforming process (1988).[14]

Figure 20.3. Comparison of the Specific Energy Consumption (SEC) in ammonia production for various countries.[a]

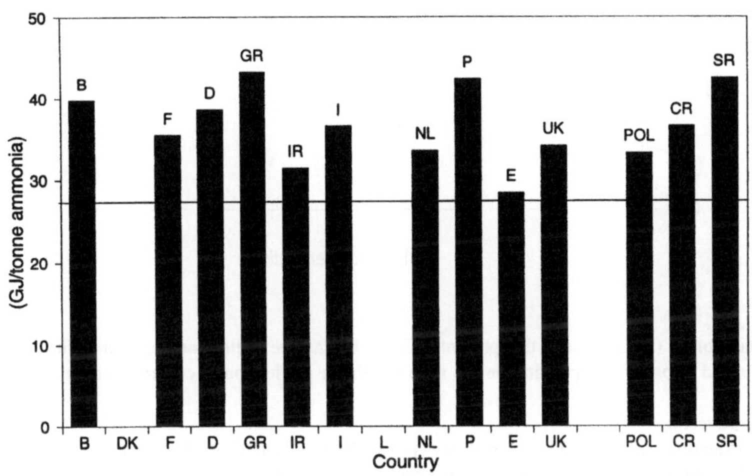

[a] The bars represent the present SEC, whereas the solid line represents the best practice currently attained with steam reforming of natural gas.

Cement production

The main energy-consuming step in the production of cement is the production of clinker. Clinker is mixed in varying amounts with other constituents in order to obtain various types of cement. Therefore, the specific energy consumption of cement is a function of the amount of clinker produced in relation to the total amount of cement produced. Clinker can be produced by a so-called wet-type or dry-type process, of which the dry-type is the more energy efficient one. For a clinker/cement ratio of

---

[14] Information for non-EU countries is retrieved from Bossenbroek, 1993; Meyers, 1993; Statistical information on the Czech Republic, 1990; Vallance, 1993.

one, the 'best practice' SEC for a dry-type process with a 5-stage preheater is assumed to be 3.3 GJ/ton.[15]

Figure 20.4. Comparison of the Specific Energy Consumption (SEC) in cement making for various countries.[a]

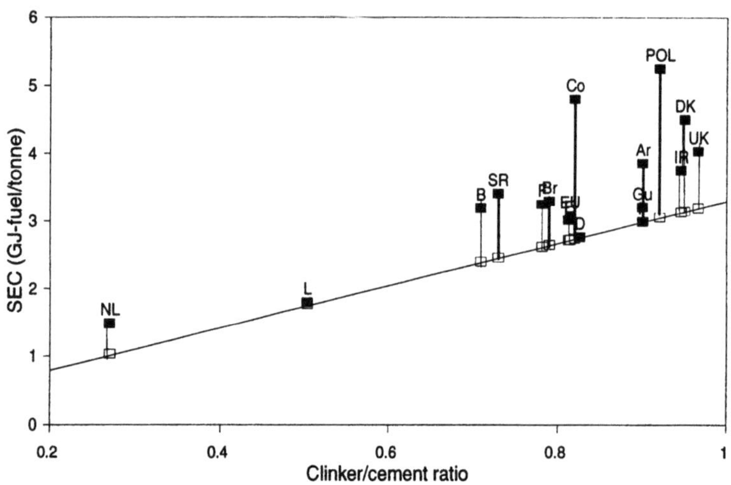

[a] The points (■) represent the present SEC, whereas the solid line represents the best practice attained at present (depending on the ratio of clinker production to cement production).

Steel production
Steel production can be based on the Basic Oxygen Furnace (BOF) or the Electric Arc Furnace (EAF) route. The BOF route uses iron ore and scrap as feedstock to primary steel production. Although through this process a higher quality of steel can be produced, the energy consumption is also higher, depending on the amount of scrap used. The 'best practice' SEC used and the efficiency improvement potentials calculated in this analysis are valid for a 10% scrap input. In the EAF route scrap is only used as feedstock for secondary steel production.

Next to feedstock and process type, the product type itself exerts influence on the specific energy consumption. In this respect a distinction can be made between slabs, hot rolled products and cold rolled products in the BOF route, and wire rods in the EAF route. Moving from slabs to cold rolled products in the BOF route results in increases in the 'best practice' SEC from 14.2 to 17.2 GJ fuel/ton and from 0.4 to 1.3 GJ/ton. These values are based on the Hoogovens steel plant in The Netherlands.[16] For the EAF route a 'best practice' SEC can be found of 1.8 GJ fuel/ton and 1.9 GJ/ton for wire rods, based on the Badische Stahlwerke plant in Germany.[17]

---

[15] Worrell et al., 1995.

[16] Worrell et al., 1995.

[17] see note 10.

The figure on the energy efficiency potential in China, roughly 33 GJ/ton steel, has been derived from a study by Ross and Feng.[18] This figure should be interpreted with care because the study did not make a clear distinction between energy used for steel production and energy used for the production of cokes. For the presentation in Figure 20.5 the figure for China found by Ross and Feng has therefore been corrected by subtracting the average energy consumption for the production of cokes.[19] Since the range of values for SEC of cokes production is large, the deviation of the present SEC from the average value could be an explanation for the rather high value for the SEC of steel production in China.

Figure 20.5. Comparison of the Specific Energy Consumption (SEC) in steel making for various countries.*

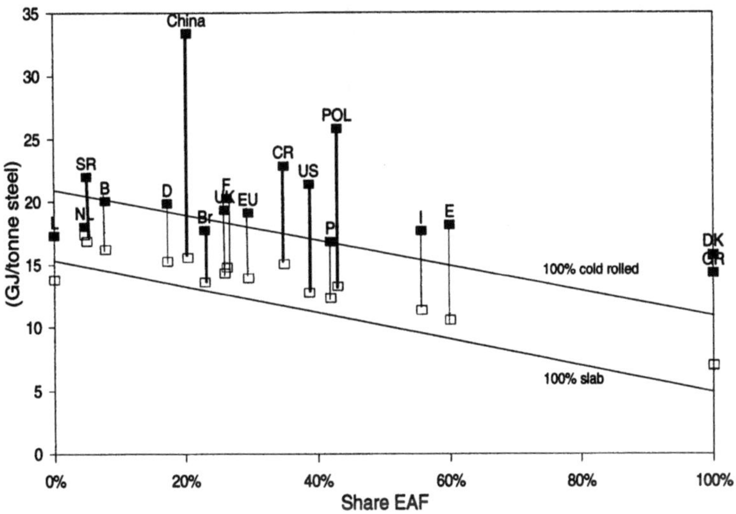

* The upper points (■) represent the present SEC, whereas the lower points (□) represent the best practice attained at present, taking into account the current mix of the feedstock and products for each specific country.

---

18 Ross and Feng, 1991.

19 We assumed a use of 553 kg-equivalent cokes/tonne pig iron. About 25% of the energy content of the cokes is used in the cokes plant. In older cokes plants this can be up to 100% of the cokes energy content.

Annex 2. Abbreviations

Ar	Argentina	f-USSR	former Sovjet Union
AUS	Australia	GR	Greece
B	Belgium	Gu	Guatemala
Br	Brazil	I	Italy
CAN	Canada	IR	Ireland
Co	Colombia	JAP	Japan
CR	Czech Republic	L	Luxembourg
D	Germany	NL	The Netherlands
DK	Denmark	POL	Poland
E	Spain	S	Sweden
E-Eur	Eastern Europe	SA	South Africa
EU	European Union	SR	Slovak Republic
F	France	UK	United Kingdom

References

Bossenbroek, E., 1993. *Energiegebruiken in Polen; trends en vergelijkingen met de EG/OECD*, (Energy Use in Poland; Trends and Comparisons with the EU/OECD), Department of Science, Technology & Society, Utrecht University, The Netherlands, November 1993.

Faaij, A., K. Blok and E. Worrell, *forthcoming. World-wide comparison of efficiency and carbon dioxide emissions of public electricity generation*, department of Science, Technology and Society, Utrecht University, The Netherlands.

Garcia del Valle, J. and A. Torres, 1990. 'Outlook of Latin American Cement Industry', in: J. Sichis, 1990. *Energy Efficiency in the Cement Industry*, Elsevier Applied Science, London/New York.

Gusbin, B., 1991. *CO_2-Study-"CRASH Programme"; Cost-effectiveness analysis of CO_2 reduction options*, Coherence, Belgium.

Howarth, R.B., L. Schipper, P.A. Duerr and S. Strom, 1991. 'Manufacturing energy use in eight OECD countries; decomposing the impacts of changes in output, industry structure and energy intensity', *Energy Economics*, April, pp. 135-42.

IEA, 1993a. *Energy Balances of OECD Countries 1990-1991*, Paris.

IEA, 1993b. *Energy Balances of non-OECD Countries 1990-1991*, Paris.

International Iron and Steel Institute, 1990. *Statistics on Energy in the Steel Industry (1990 Update)*, Brussels.

International Iron and Steel Institute, 1992. *Steel Statistical Yearbook 1992*, Brussels.

Kram, T., 1993. *National energy options for reducing CO_2 emissions, volume I: The international connection*, ETSAP, ECN, Petten.

Meyers, S., L. Schipper, J. Salay, A. Gromadzinski, E. Hillie, P. Kaleta, M. Kumanwoski, J. Maron, J. Norwisz and S. Pasierb, 1993. *Energy Use in Poland, 1970-1991: Sectoral Analysis and International Comparison*, LBL, Berkeley, July 1993.

Nilsson, L.J., 1993. 'Energy Intensity Trends in 31 Industrial and Developing Countries 1950-1988', *Energy*, 18, pp. 309-22.

OECD, 1991. *The State of the Environment*, Paris.

Ross, M., L. Feng, 1991. 'The Energy Efficiency of the Steel Industry in China', *Energy*, 16, pp. 833-48.

Schipper, L., R.B. Meyers, R. Howarth and R. Steiner, 1992. *Energy Efficiency and Human Activity: Past Trends, Future Prospects*, Cambridge University Press, Cambridge USA.

Schipper, L., R. Steiner, M.J. Figueroa and K. Dolan, 1993. 'Fuel prices and economy; factors affecting land travel', *Transport Policy*, 1, pp. 6-20.

Schipper, L., S. Meyers and R. Howarth, 1993. 'Energy intensities in OECD countries: an historical analysis', *International Journal of Global Energy Issues*, 5, pp. 76-89.

Statistical information on the Czech Republic, Slovak Republic and Poland, 1990.

Swisher, J. and A. Villavicencio, 1992. *The UNEP greenhouse gas abatement costing study*, UNEP Collaborating centre on Energy and Environment, Risø, Denmark.

Vallance, B., 1993. *East-West Comparisons of Energy Efficiency in Energy Intensive Industries*, Symposium on Energy Efficiency and Economic Transition in Central and Eastern Europe, Paris, 25-28 May 1993.

Worrell, E., R.F.A. Cuelenaere, K. Blok and W.C. Turkenburg, 1995 *(forthcoming)*, 'Energy Consumption by Industrial Processes in the European Union', *Energy*.

21. Joint Implementation in the Cement Industry

JAAP JANSEN and FRANK VAN DER VLEUTEN*

1 Background

In the framework of The Netherlands' National Research Programme on Global Air Pollution and Climate Change (NOP-MLK) a collaborative study is currently being carried out to assess strategies and instruments to promote industrial energy efficiency in developing countries. In the developing world almost half of the final use of commercial energy is related to industrial activities. The study particularly concentrates on the following subsectors: iron and steel; aluminium; pulp and paper; and cement. It provides the background of this chapter, which will discuss the opportunities for JI in the cement industry.

2 The Cement Industry: an Overview

Cement is a binding component which primarily derives its importance from its main end-use, concrete. Concrete is used in the strategically important construction sector, which has strong linkages to industry, transport, utilities, services and housing. Because of its high transportation costs and its perishable nature, cement is mostly produced in the proximity of the market and has only a modest share in international trade. Therefore, cement plants can be found in most countries in the world. As the cement industry has been considered to be of strategic importance, many governments of developing countries are still considerably involved in the ownership and management of their cement plants. However, in recent years the drive towards privatization is gathering strength.

Cement is produced and consumed throughout the world. Although per capita consumption and production levels in developing countries are still relatively low, by now the latter accounts for roughly half of the world's cement production (571.8 mln. tons out of 1147.3 mln. tons in 1990, see Table 21.1). Statistical analysis shows that per capita cement consumption is positively correlated with:

---

* The authors are associated with the Netherlands' Energy Research Foundation ECN, Petten, The Netherlands; this chapter is based on a study jointly conducted by the energy research institutes ECN (Petten, the Netherlands), ENDA (Dakar, Senegal), IEI (Rio de Janeiro, Brazil), and TERI (New Delhi, India).

C. J. Jepma (ed.), The Feasibility of Joint Implementation, 299–313.

- the level of economic development;
- long-term economic growth performance; and
- upward swings of the business cycle.

Table 21.1. World cement production and consumption in 1990.

	cement production (mln. tons)	cement consumption (Kg/cap.)
OECD	387.6	465
North America	83.4	324
Europe	211.9	498
Pacific	92.3	638
Africa	55.1	94
North Africa	35.1	317
Sub Saharan Africa	20.0	44
Rep. South Africa	7.9	214
Asia (excl. Japan)	432.5	148
China	209.7	180
India	47.3	57
Other Asia	175.5	190
Latin America & the Caribbean	84.2	166
South America	48.9	172
Central America	27.2	214
Caribbean Islands	8.1	261
Transition Economies	187.9	441
Former USSR	137.3	479
Others	50.6	352
World	1147.3	217

Source: Cembureau, 1992.

3 Technology and Energy Use

Cement is mainly produced in large plants from raw materials containing $CaCO_3$ (generally limestone), SiO_2, Fe_2O_3, and Al_2O_3. The raw materials are first crushed, then mixed, and finely grounded into what is called *raw meal*, which is fed into a kiln and then gradually heated to approximately 1600°C. When all moisture has been evaporated, $CaCO_3$ is decomposed into CaO and CO_2. In the highest temperature zone, clinker, the main intermediate product in cement production, is formed through a sintering process. The clinker is subsequently interground with 5% gypsum to give ordinary *Portland cement.* By intergrinding clinker with high amounts of secondary constituents, blended cements can be produced, such as *blast furnace slag cement, fly ash cement,* and *pozzolanic cement.*

This process of cement production is quite energy-intensive. Depending on local prices, energy directly accounts for roughly 30% to 40% of total production costs. In addition, a non-negligible amount of energy is used for transportation to deliver cement to the final user, even though the cement industry is often situated relatively close to the market. Most energy is used in the phase of producing clinker. In the kiln, fuel is burned to create the high temperatures required for the endothermal decarbonization reaction. Broadly, two different kiln technologies can be discerned: the outdated *wet* process (of which specific energy consumption in OECD countries currently amounts to 5,500 MJ/t clinker); and the considerably more efficient *dry* process (3,600 MJ/t clinker in the OECD countries). Apart from kiln efficiency, the total fuel use per ton of cement is determined by the amount of secondary constituents added in clinker grinding. The cement industry furthermore consumes a considerable amount of electricity in: crushing and grinding mills; fan and kiln drives; conveying; and packaging plants. Generally, cement production requires roughly between 90 and 160 kWh of electricity per ton of cement, depending on the grinding technology used, but also on the material and the kiln technology used.

An overview of specific energy consumption (average energy use per unit of output) in the cement industry is given in Table 21.2. The table indicates that, by and large, developing countries use less efficient kiln technology and less secondary constituents. Moreover, the technology embodied in plants in developing countries tends to operate less efficiently than in plants in OECD countries. In industrialized countries, the cement sector generally accounts for between 1% and 6% of the total commercial energy use; in developing countries - especially in some smaller countries - the corresponding share is considerably higher, *i.e.* between 3% and 10%. As a result, in most developing countries, the cement industry ranks among the largest industrial energy users.

Table 21.2. Overview of specific energy consumption in the cement industry by region (1990).

	share of wet process (%)	secondary constituents (%)	specific heat cons. (MJ/t clinker)	electricity use (kWh/t cement)
OECD				
USA[a]	35	na	4,250	138
Germany	5	21	3,610	104
The Netherlands	0	46	3,710	96
Japan	3	na	2,970	103
Africa				
Egypt	53	na	4,840	90
Nigeria	43	na	na	na
Zimbabwe	0	na	na	na
Asia				
India[b]	40	na	4,780	126
China[c]	61	na	5,510	106
Indonesia	20	5	3,766	na
Latin America				
Costa Rica	0	7	3,580	132
Argentina	35	10	4,290	128
Brazil	22	21	4,160	125
Colombia	80	na	5,860	122
Transition econ.				
former USSR	81	28	6,301	107
Poland	60	18	5,565	na

a Data for 1985.
b Not included estimated number of 180 mini cement plants (6% of total capacity).
c Not included estimated number of 6,000 mini cement plants (85% of total capacity).
na not available
Source: Vleuten, Van der, 1994.

4 Greenhouse Gas (GHG) Emissions

In cement production carbon dioxide (CO_2) is a byproduct of pyroprocessing originating from the combustion of kiln fuel and the decarbonization reaction:

$$CaCO_3 \rightarrow CaO + CO_2 \ - 1620 \text{ kJ}$$

Furthermore, demand for electricity brings about CO_2 emission during thermal electricity generation. Table 21.3 shows the estimated contributions of the different sources to the total CO_2 emission by the cement industry in five groups of countries.

Table 21.3. Estimated global CO_2 emissions associated with the production of cement, by region and source, 1990.

	CO_2 emissions from cement production by source (MtC_{eq})			total CO_2 from cement production (MtC_{eq})	total man-made CO_2 emissions (MtC_{eq})	share cement in man-made CO_2 emissions (%)
	fuel	power	decarb.			
OECD	24	5	41	70	3,011	2.3
Africa	5	1	7	12	167	7.2
Asia (excl.Japan)	44	6	56	107	1,286	8.3
Latin America & Car.	7	1	10	17	270	6.3
Transition Economies	17	3	19	39	1,279	3.0
World	97	16	133	245	6,013	4.1

Note: Total man-made CO_2 emissions shown in this table include energy related emissions and additional emissions from cement manufacturing (calcination process), but do not include net emission additions due to a reduction of 'sinks' (deforestation, land-use change).

Sources: Vleuten, Van der, 1994; Buskens et al., 1994.

Due to the hefty emissions from decarbonization, which is unique to cement production and in fact accounts for a level of emissions that is even higher than in energy production, the cement industry is one of the foremost sources of worldwide GHG emissions. Moreover, in many developing countries the cement industry, due to its relatively large share in industrial energy use, is the largest contributor to CO_2 emissions. From the figures presented in Table 21.3 it can be inferred that over 4% of global anthropogenic CO_2 emissions (excluding land-use change) in 1990 can be attributed to the cement industry.[1] The last column of Table 21.3 furthermore reveals that the share of the cement industry in developing countries in total CO_2 emissions is far above the world average. Compared with the OECD countries, this outcome relates to high levels of specific energy consumption, high emissions per unit of power generated (especially coal-rich China and India) and little use of clinker substitutes.

[1] It should be noted that this percentage does not include CO_2 emissions generated by the construction of cement plants and by haulage of cement from the plant to distribution centres and final users.

5 CO_2 Abatement Measures

With respect to the cement industry, many 'no regrets' options, at least from a national-economic perspective, exist in developing countries and in countries with economies in transition. An easy capitalization of these opportunities is often inhibited by the oligopolistic nature of the market for cement production technology and cement production itself as well as a range of institutional factors within many developing countries. Such factors include the difficult access to relevant information and the pursuance of low-performing policies at the macroeconomic and the sectoral level. For instance, protectionist trade policies with strong sectoral biases and overregulated banking, cost-plus pricing practices and subsidization of energy are bound to result in foreign exchange scarcities, mounting inefficiencies and an under-development of local capital markets.

At the plant level there are four areas at which useful measures to abate CO_2 emissions can be taken:

- process management;
- process technology;
- raw material inputs; and
- energy inputs.

Each of these will be briefly discussed below.[2]

5.1 Process management

In most cement plants, and especially in developing countries, there is considerable scope for the tightening of operation and maintenance procedures. Improving operation efficiencies through upgraded 'housekeeping' would cost relatively little and could lead - through energy savings up to 20% and reduced material losses - to substantial reductions in CO_2 emissions.

A major reason for suboptimal operation and maintenance of cement plants is that in developing countries the technology for cement production is often not appropriate for the existing local production facilities and insufficiently fits the prevailing conditions. A major underlying factor for this is the inadequacy of technology transfers. New plants or modifications to existing plants are often purchased on a turnkey basis, which mostly involve large-scale transfers for state-of-the-art technology. Regularly,

[2] Vleuten, Van der, 1994.

in advance of these transfers neither the buyer nor the supplier sufficiently considers the necessary critical adjustments to overcome specific local constraints. State-of-the-art technology tends to dazzle buyers and funders whilst the suppliers, of which there are only a handful, have no real incentive to look beyond the sophisticated markets for which their technologies have primarily been developed. Often turnkey projects are handed over without an adequate backup in terms of:

- the provision of relevant information, the so-called 'infoware';
- the skills and experience in the relevant market, *i.e.* 'humanware'; and
- the internal organization of the industry and the arrangements and linkages with the plant environment, i.e 'orgaware'.

By providing the cement industry with these less tangible but essential inputs, great strides forward can be made in terms of efficiency improvements and environmental sustainability at relatively low (gross) costs.

5.2 Process technology

Installing efficient process technology, such as efficient kilns and efficient grinding mills, could considerably increase energy efficiency. This may imply small retrofits as well as total process conversions. An example of the latter is the replacement of an inefficient wet production process by a state-of-the-art dry process, which would cut fuel use and fuel related CO_2 emissions by almost 50%. Efforts to improve the efficiency of cement production processes have thus far persistently tended to concentrate on technology measures. However, in order to further improve energy efficiency it is important to keep in mind that the use of more efficient process technology places strict requirements on the available human, financial and physical resources.

5.3 Raw material inputs

As mentioned earlier, the production of clinker accounts for most of the CO_2 emissions in the cement industry. In the production of cement used for general purposes, considerable abatement is possible by replacing a large proportion of clinker by secondary raw material without significantly affecting the cement quality. In this respect blast furnace slag offers the highest substitution potential, as it could be used to replace up to 80% of the clinker, but also fly ash and natural pozzolanas (volcanic soils) have a considerable potential. The reasons, however, why the use of these additives is not yet widespread are related to: an initial reluctance by the industry to pursue this road; a consequent general lack of information about experience with these materials; and underdeveloped collection and distribution channels.

5.4 Energy inputs

Cement kilns often produce considerable amounts of waste heat, which could be reintroduced into the process, either directly for drying or preheating purposes, or indirectly by co-generation. In case such a recycling is not possible, various opportunities remain for using waste heat for 'over the fence' purposes. Evidently, improving the use of waste heat offers a considerable energy conservation and emission abatement potential.

Furthermore, cement kilns have inherent properties which make them suitable for burning waste materials. At least theoretically, cement kilns could operate efficiently on waste alone and thus alleviate the increasing need for very costly incineration plants and increasingly unacceptable waste disposal sites. Clearly, the question whether using waste material as a fuel in the production of cement would be suitable in terms of energy conservation and pollution abatement would first need to be comprehensively addressed. When introducing 'waste' fuels, similar barriers have to be surmounted as for the introduction of secondary raw materials.

Regional CO_2 abatement potentials have recently been estimated by Van der Vleuten.[3] The estimates were based on the assumptions:

- that all present technology in cement production would be replaced by well operating state-of-the-art technology;[4] and
- that throughout the world, 50% of the raw materials input in cement production would consist of secondary raw materials.[5]

The resulting estimated CO_2 abatement potentials are summarized in Table 21.4.

---

3 Vleuten, Van der, 1994.

4 Using Japanese cement production data.

5 At present, this is approximately the case in The Netherlands.

Table 21.4. Estimated CO_2 abatement potentials.

Region	State-of-the-Art technology (%)	50% Additives (%)	Both (%)	Both (MtCeq)
OECD	7	33	38	27
Africa	12	41	48	6
Asia (excl. Japan)	16	44	53	57
Latin America & Caribbean	13	39	47	8
Transition Economies	21	30	45	18
World	14	38	47	115

Sources: Vleuten, Van der, 1994; Buskens et al., 1994.

6 Possibilities for international cooperation

In the cement industry in OECD countries there is considerable experience with knowledge of and skills in:

- optimizing process operations;
- upgrading process technology; and
- using secondary raw and/or waste materials.

If harnessed in a proper international cooperation framework this experience could significantly contribute to the achievement of global GHG abatement at low costs and to the realization of the desired improvements in the cement industry in developing countries. In terms of overall productivity performance and of energy efficiency in particular these improvements can be achieved at three points:

- building new plants, mounted with appropriate technology;
- appropriate improvements in existing production facilities; and
- improvements in sector-wide facilities.

6.1 Building new plants

Most cement plants serve large, relatively sophisticated markets and should therefore be large-scale in order to exploit economies of scale. At present, the competitiveness of these markets varies, largely reflecting differences in the vigour with which privatization programmes are being pursued. Over time, however, they would seem to be set to converge. The crucial issue here is how to effectively transfer and adapt available state-of-the-art technology so that it can overcome location-specific constraints in human and physical resource availabilities.

International development cooperation may have to play a useful role in this process. New forms of development cooperation should help to supplant the conventional relationships between the supplier of the technology and/or the funds on the one hand and the buyer (recipient) on the other, by effective two-way partnerships to ensure that:

- at the origin of the technology transfer appropriate attention is paid to location-specific bottlenecks; and
- at the destination of the transfer, *i.e.* plant location, ways and means are available to adequately assess the prevailing constraints and negotiate effective solutions.

In the absence of well-developed, integrated cement markets, existing intermediate technologies - such as the low capital, high labour-intensive mini-cement technology as currently applied in China and India - may still have a useful role to play in developing countries for a long time to come. In this respect for the small, scattered and relatively inaccessible markets that still exist in many developing countries, international development cooperation can contribute to improve the productivity, the quality and the energy-efficiency of these intermediate technologies.

6.2 Existing production facilities

A wide range of modalities of commercial and development cooperation are eligible for consideration to upgrade existing plants. International cooperation between private agents can proceed through direct foreign investments or through delivery of equipment and technical assistance on a commercial basis, *e.g.* in the form of management contracts. The current wave of partial or complete privatizations of parastatal companies provides ample opportunity for international cooperation between private entities.

In this respect interventions by 'bridging agents' (investment-facilitating institutions) can make important contributions, especially in areas that are of marginal interest to the transnational players in the cement industry. One can think of organizations that mediate in technical assistance services, *e.g.* multilateral organizations such as UNIDO and the European Community Investment Partners programme (ECIP) of the EU.

Also multilateral, bilateral and national development banks may offer interesting non-banking investment promotion services including the relevant consulting services and management assistance. Multilateral development banks with non-banking activities in the cement industry include: the World Bank and affiliated institutes (especially IFC); the Islamic Development Bank; and regional development banks such as the Asian Development Bank. An example of a bilateral development bank is the Dutch Finance for Development Company (FMO), which offers a wide range of non-banking facilities by way of its Investment Promotion and Technical Assistance (IPTA) programme. National development banks in developing countries - often acting as operating agent or on-lending channel of international development cooperation agencies - may also offer relevant non-banking services for the upgrading of existing cement plants.

6.3 Sector-wide facilities

The establishment of new plants and the upgrading of existing plants can be optimized by: improving the sector-wide availability of relevant information; improving the opportunities for human resources development; and providing technical assistance. Such interventions to create sector-wide facilities would include the sponsoring of cooperation between or twinning of branch organizations, as well as supporting relevant multilateral activities.

At the branch level, national or regional (inter-country) industrial branch associations, such as the European Cembureau, may be considered for participation in international cooperation activities. In the South regional branch cooperation tends to be weakly developed. It could be considered, for instance, to include the cement sector in the Preferential Trade Area (PTA) industrial umbrella organization of anglophone Africa or, alternatively, in African national branch organizations. An example of relevant activities by multilateral development agencies might be the unbiased brokerage by UNIDO in the participation of experts from the African cement industry in traineeships and on-the-job-training in Europe.

In Table 21.5 an overview is presented of major options for international cooperation in the cement sector in developing countries. In principle, this cooperation can be set up on a commercial basis, either in the framework of regular development cooperation (ODA) or through JI. Because commercial private sector cooperation will most likely focus on countries with well-established markets or with good growth perspectives (for example, the middle-income developing countries and economies in transition), regular development cooperation and JI should pay special attention to areas which are not covered by private-sector activities (notably including the low-income developing countries).[6] However, since global cost-effectiveness is an overriding criterion for JI, options with cost-effective additionality with respect to GHG abatement in middle-income developing countries should certainly not be excluded.

Table 21.5. Broad categories of options for international cooperation in the cement sector.

option/region	commercial private-sector cooperation	official development assistance	JI activities
New plants			
Joint venture North-South: OECD technology	2	1	1, 2
Joint venture South-South: mini-plant technology		1	1
Existing plants			
Management assistance	2	1	1, 2
Specialized technical assistance	2	1	1, 2
Small retrofits	2	1	1, 2
Profound process changes	2		2
Sector-wide facilities			
Twinning branch organizations		1	1, 2
Human resource development	2	1	1, 2

1 Eligible for low-income countries with modest economic growth prospects.
2 Eligible for middle-income countries with prospects for high economic growth.

By and large, Africa faces stagnating economic development and political instability. The size of the cement industry in this region is therefore still relatively modest and the prospects, from the perspective of transnational cement companies, are unclear as yet resulting in rather high commercial risks. In Africa, many cement plants face low rates of capital utilization, caused by unplanned production shutdowns due to internal factors (insufficient mastery of the technology, poor logistics planning, bad maintenance) as well as to external factors (stagnating demand and, for example, unforeseen transportation bottlenecks, power cuts, lack of foreign exchange to acquire essential spare parts, problems in getting vital foreign spare parts through customs, *etc.*). These

[6] Compare in this respect FCCC, Art. 4.9: 'The Parties shall take full account of the specific needs and special situations of the least developed countries in their actions with regard to funding and transfer of technology.'

factors and the commercial risks have created high barriers to attract investments by transnational corporations.

Thus far, 'spontaneous' international private-sector cooperation to improve the performance of African cement plants has been limited. Therefore, the African cement industry is an interesting candidate for intervention through international development cooperation. In principle, many opportunities exist for cost-effective GHG abatement either through ODA or through JI. However, these activities should be well-targeted and should fit into ongoing structural adjustment programmes. In this respect it can be argued that the effectiveness of development cooperation activities in a country would decline for the following reasons: a large government interference in the daily management of the industry; a strongly protectionist trade and industrial policy; and high energy subsidies.

With respect to other developing regions it can be concluded that in the rapidly growing South East Asian countries' cement plants are attractive targets for direct foreign investments. Also in other (low-income) Asian countries considerable scope for complementary development cooperation and JI seems to exist. For Latin America and the more advanced economies in transition, more or less the same applies as for South East Asia. In general terms, one may infer from the above analysis that in the cement industry in larger developing countries with good prospects for economic growth, the eligibility for development cooperation and JI is less than in developing countries facing a slow economic growth.

7 Conclusions

- In the cement industry, in particular in the developing world, ample opportunities exist for GHG abatement on a 'win-win' basis from the perspective of the recipient as well as the donor party.
- The cement industry is a suitable candidate for targeting JI, because of:
    - the large scope for GHG-emissions reductions that exist in the cement industry, in particular in developing countries and economies in transition;
    - the compatibility with general development priorities of developing countries (stimulation of economic development through implementing 'win-win' options and strengthening of the 'strategic sector');
    - the transparency of the alternative production processes and homogeneous products, which facilitates the definition of the 'baseline'; and
    - the highly localized emissions, making monitoring more practicable.

- In particular in Africa and in the low-income Asian and Latin American countries there seems to exist ample scope for JI as a catalyst to international cooperation between Southern and Northern private-sector agents, thereby capitalizing on opportunities for cost-effective GHG abatement.
- The potential for successfully improving general levels of productivity and energy efficiency, and for market-triggered dissemination of GHG-extensive technology, is largest in those countries that have already initiated apparent policy steps to improve economic management at the macro-level as well as at the sector-level. In identifying and screening potential JI projects, this aspect should be given significant weight.
- So far, some OECD country governments (such as The Netherlands and Canada) have been fairly successful with the environmental policy instrument of voluntary multi-annual agreements - the so-called 'covenants' - between private-sector companies and governments on energy-efficiency and other environmental objectives. This holds, among other sectors, notably for The Netherlands' cement industry. Due allowance for partly meeting the targets in the covenants abroad could generate more support for JI from the private sector in industrialized countries. To further stimulate the involvement of private-sector agents in JI as projects in developing countries with slow economic growth but which nevertheless have initiated sound macroeconomic policies, additional incentives such as, in particular, subsidies should be considered.

References

Buskens, V.W., and J.C. Jansen, 1994. *Industrial Energy Demand and CO_2 Emissions in Developing Countries in Global Perspective*, NOP/MLK Project Strategies and Instruments to Promote Energy Efficiency in Developing Countries, Working Paper no. 1, ECN, Petten (report: ECN-C-94-039).

Cembureau, 1992. *World Statistical Review no.14/1989-1990-1991*, Cimeurope, Brussels.

Kuik O., P. Peters and N. Schrijver (eds.), 1994. *Joint Implementation to Curb Climate Change*, Kluwer Academic Publishers, Dordrecht/Boston/London.

Lang T.A., 1994. *Energy Savings Potential in the Cement Industry and Special Activities of the "Holderbank" Group*, paper presented at the IEA/US DOE international workshop 'Industrial Energy Efficiency: Policies and Programmes', Washington, DC, May 26-27, 1994.

Vleuten, F.P. van der, 1994. *Cement in Development: Energy and Environment*, NOP/MLK Project Strategies and Instruments to Promote Energy Efficiency in Developing Countries, Working Paper no. 4, ECN, Petten (report: ECN-C-94-035).

World Resources Institute, 1992. *World Resources 1992-93*, Oxford University Press, New York/Oxford.

22. The Economics of Managing Carbon via Forestry

An Assessment of Existing Studies

ROGER A. SEDJO*

1 Introduction

There is vast literature examining the carbon implications of various types of land uses including land with and without forests, tree planting, other biomass growth, *etc.* A recent book quantified the potential for managing terrestrial systems to sequester carbon (*e.g.*, tree planting, or utilizing biomass fuels).[1] However, with few exceptions the articles in this book and other assessments lack a discussion of financial and/or cost considerations. This chapter is directed at examining what has been reported in the literature. The costs and/or the benefits of the various land use and tree planting schemes are placed in financial or economic terms.

This chapter will: review literature dealing with the costs of sequestering carbon as a vehicle to mitigating global warming; assess in general the adequacies and deficiencies of the literature; and suggest future directions. Mitigation policies involving the forest sector fall into one of four categories: those that increase the standing inventory of forest biomass and therefore sequester carbon; those that increase the storage of carbon in long-lived forest products; those that substitute wood products for other materials that emits more carbon;[2] and those that utilize biomass energy crops for fuel.[3] Mitigation policies using forests have been estimated to reduce by as much as 80% the overall costs of stabilizing US carbon emissions.[4]

2 Conversion of Marginal Crop and Pasture Lands into Trees and Forests

Much of the early literature examined the question of the relationship between forest area and/or biomass and the amount of carbon sequestered in the forest, providing

---

* The author is senior Fellow associated with Resources for the Future.

1 Wisniewski and Sampson, 1993.

2 Binkley and Van Kooten, 1993.

3 Wright and Hughes, 1993; Scurlock, et al., 1993.

4 Richards, et al., 1993.

C. J. Jepma (ed.), The Feasibility of Joint Implementation, 315–327.

valuable information on methods to increase the standing inventory of forest biomass. Marland (1987) pulled together much of the earlier literature on carbon sequestration by forests and soils, and estimated the relationship between various types of forest conditions and the amount of carbon sequestered. Sedjo and Solomon (1989) made what appears to be the first estimate of the costs of sequestering major amounts of carbon by forest establishment, estimating that it would take about 465 mln. ha of new forest at a cost of about \$186-\$372 bln. to sequester enough carbon on an annual basis to offset the current net CO_2 emission at that time. Their estimate of cost was based upon estimates of average land prices in the temperate and tropical regions as well as estimates of the average costs of establishing forest plantations on these lands. Using average costs, they obtained crude point estimates that did not take into consideration increasing costs that are likely to occur as forest plantation establishment moves from land with few alternative uses to more productive land. Darmstadter and Plantinga (1990) derived an estimated cost of \$2-\$4/tC from the Sedjo and Solomon gross estimate.

A widely cited, but extreme, early estimate is that of Nordhaus (1990) in which the cost of the removal of carbon in the USA were estimated at \$100/tC. This cost estimate is almost surely too high, at least as an estimate of the current marginal costs of sequestering carbon by tree planting. The high Nordhaus cost estimate has been challenged by Cline (1992). The result is strongly affected by Nordhaus' low estimate of carbon sequestered per acre of forest, about 0.4 tons, which is less than one tenth of the estimate used by Sedjo and Solomon (1988) and Moulton and Richards (1990). This low estimate of carbon sequestered appears to reflect:

a) failure to distinguish between industrial wood and forest woody biomass (which includes branches, bark and other wood wastes);
b) the much more rapid growth rate of a young forest plantation that would be developed for carbon sequestration compared with the low average growth of a mature forest; and
c) the absence of an estimate for the effects on carbon of the buildup of detrital material and soil carbon related to forest growth.

A number of recent studies have also examined the question of the costs of sequestering carbon globally. For example, Winjum et al. (1992) estimated that the costs of sequestering carbon on a massive worldwide scale would range from less than \$1 to about \$8/tC sequestered. Dixon et al. (1993) estimated the cost of carbon sequestration to vary from \$1 to \$30/tC for projects proposed by US and European utilities as carbon offsets outside of their regions. However, many of these studies often did not include all costs. For example, the Winjum et al. (1992) study treated land required for establishing carbon plantations as free, therefore implying zero opportunity costs. Furthermore, although the study discounted costs, it made no attempt to discount the carbon benefits. In the Dixon et al. (1993) cases, some of the costs do not appear to be realistic, such as those of land and voluntary labour costs.

In fact, many projects will also have benefits other than those associated with the carbon sequestration. The marginal cost of the carbon activity should be estimated as the total costs of the project less the non-carbon benefits, all properly discounted. This calculation has not been attempted. However, in fairness, the noncarbon benefits are difficult to determine in most cases.

Other studies have been done recently to estimate the costs of sequestering carbon via specific forestry activities, usually in developing countries. Some of these are listed in Table 22.1.

Table 22.1. Costs of sequestering carbon through forest projects: some selected cases ($/tC).

Source	Tropical		Temperate Plantation	Boreal	
	Agroforestry	Plantation		Plantation	Protection
Andrasko (1993)	3-5	3-6	0-2		
Dixon et al. (1993)	4-16	6-60	2-50	3-27	1-4
Krankina and Dixon (1993)			1-7	1-8	1-3
Houghton et al. (1991)	3-12	4-37			

Source: Adapted from Dixon et al., 1993.

As indicated above, a host of studies have suggested relatively low carbon sequestering costs through tree planting; in many cases under $10/tC and rarely over $30/tC. Other studies with similar results stressing various aspects of the problem include: Trexler et al. (1989); Swisher (1991); Winjum and Lewis (1993); and Faeth et al. (1993). Another early US study which included alternative options is that of Dudek and LeBlanc (1990) who examined the use of tree planting under the Conservation Reserve Program (CRP) to sequester carbon. They estimate this option as costing from $26 to $43/tC removed from the atmosphere and for other options the costs are estimated at from $4.40 to almost $240/tC.

In a recent study Turner et al. (1993) presented a range of estimates of the cost of forest establishment per ton of carbon sequestered estimated by forest type and the mode of forest regeneration. These are presented in Table 22.2 below. These cost estimates, by region and forest regeneration mode, are quite consistent with the cost range of Table 22.1.

Table 22.2. Establishment costs of cost-efficient practices.

Forest type/practices	Median	Median $/ha.[a]
Boreal:		
natural regeneration	5 (4-11)	93 (83-126)
reforestation	8 (3-27)	324 (127-455)
Temperate:		
natural regeneration	1 (<1-1)	9 (9-100)
afforestation	2 (<1-5)	259 (41-444)
reforestation	6 (3-29)	357 (257-911)
Tropical:		
natural regeneration	1 (<1-2)	178 (106-238)
agroforestry	5 (2-11)	454 (254-699)
reforestation	7 (3-26)	450 (303-1183)

[a] The numbers in parentheses are interquartile ranges (middle 50% of observations).
Source: Drawn from Turner et al., 1993.

The Sedjo and Solomon (1989) work and much of the other work cited above provided single point estimates, and thus do not accurately represent the very real problem of rising costs that are associated with involving large land areas. For large projects, the marginal costs would be expected to rise reflecting both increasing land opportunity costs and varying land productivity for tree growing. In most of the previous studies, a point estimate was justified since the focus was only on a single project. However, point estimates are not appropriate when examining policy options for projects of the scale needed to offset large amounts of carbon.

In addition, many early cost estimates ignored the use of a discount rate in their calculations or used crude approximations involving averages. However, since the costs of most of the forestry investments are incurred before the carbon sequestration benefits, some type of discounting procedure is in order and crude approximations may not provide accurate estimates.

Recently, a number of somewhat more sophisticated studies have begun to appear. These include Moulton and Richards (1990), Adams et al. (1992), and Parks and Hardier (1992). These studies have refined the approach of estimating the costs of establishing carbon sequestering tree plantations in three ways.

First, they estimated a cost function, not a point, that would indicate the rise in costs associated with large-scale tree planting. In doing this they recognized that large scale investments in forests to sequester carbon would face rising costs as forests moved from lands with low productivity and/or low opportunity costs to lands with higher productivity and/or opportunity cost. Second, they refined the tree plantation establishment costs estimate recognizing differences associated with location and site considerations. Third, they built discounting procedures into the methodology.

Moulton and Richards (1990) developed the first cost function for the USA by systematically building an upward sloping marginal cost curve that ranges from \$5-\$10/tC for the first 60 mln. tons increasing to \$40/tC as annual sequestration rises to 800 mln. tons and total land area involved approaches 340 mln. acres (about 130 mln. ha). Thus, the estimates captured the reality that under current conditions the marginal costs of sequestering a ton of carbon in the USA is modest, less than \$10/tC, and rises gradually to about \$40/tC when large amounts of land, which require higher opportunity costs for land to be utilized, are involved.

The estimates of Moulton and Richards (1990) are roughly equivalent to the point estimate of Sedjo and Solomon (1988) and others for low levels of forest plantations but change dramatically for higher levels of tree planting, since they incorporate rising marginal costs as lands with higher opportunity costs are brought into forest plantation production. Both approaches used roughly the same assumption regarding the amount of carbon that would be sequestered per land area planted, *i.e.*, about six tons per ha per year. The work of Adams et al. (1992) is similar to that of Moulton and Richards (1990) in that it estimates a rising cost function. The study estimates of sequestration costs are somewhat higher than Moulton and Richards (1990), being in the range of \$16.30 for low levels of carbon sequestration to \$62/tC for a US programme that sequesters about one-half of US carbon emissions. In a recent paper Adams et al. (1992) make modest upward revisions in the estimates but they are essentially in the same range.

A recent study that generates results somewhat consistent with the high cost estimates of Nordhaus (1990) is that of Parks and Hardie (1992). Their study also estimates a rising marginal cost curve for carbon based on a hypothetical programme developed along the lines of the US Conservation Reserve Program (CRP). This programme is designed to encourage: the planting of grass or trees on eligible (environmentally fragile) farmlands to reduce both total agricultural output (which is in surplus), and to reduce environmental damages generated by practicing agriculture on these lands, *e.g.*, wind and water erosion, stream sedimentation; and to improve fish and wildlife habitat. Enrolment for this programme is based on bids submitted to the government that are to offset establishment and opportunity costs of devoting lands to various conservation practices, including planting trees. The government makes annual rental payments to the owner with the owner responsible for half the establishment costs and all of the costs of maintaining the chosen cover vegetation.

Parks and Hardie (1992) derived a marginal cost schedule for carbon sequestered in trees planted on marginal agricultural lands in the USA. For the entire hypothetical programme, the study estimates the average cost of sequestering carbon range from $108 to $298 per annual ton. The lower levels of carbon sequestered in their study from those of Moulton and Richards (1990) are attributed by Parks and Hardie (1990) as due to their: failure to include soil and 'under-storey carbon'; use of lower forest yields; capitalization procedures, which spread costs over 40 years; and lower opportunity costs. The lower opportunity costs of the Moulton and Richards study reflect the fact that their study includes the option of establishing improved forests on lands which have low opportunity costs. The three sets of cost estimates appear in Table 22.3.

It is clear that both the methodology and the empirical estimates are still in a state amenable to further revision. It is probably an accurate generalization to state that the newer research is tending to find a somewhat steeper increase in costs than did the early studies. As noted, the Adams et al. (1992) estimates have very recently been revised. Similarly, the Moulton and Richards (1990) estimates are being refined by incorporating dynamic patterns of carbon uptake.[5] This update addresses the inter-temporal issue by moving away from the use of average forest carbon uptakes to specific species and age related forest carbon uptakes. In addition, the analysis uses revised carbon estimates and builds in greater detail on the opportunity costs of land in developing a long-term marginal cost curve for carbon uptake. In general, the new results show a much steeper rise than the earlier Moulton and Richards (1990) work,

[5] Richards et al., 1993.

with the costs per ton roughly doubling, from about $30 to $60 per ton, for large annual uptakes.

Table 22.3. Estimates of Costs of Carbon Sequestered by Tree Planting: Some Comparative Results for the USA.

Study	Total Carbon Sequestered (in mln. tons per year)			
	140	280	420	700
	$/ton			
Moulton/Richards (1990)	16.57	20.69	23.24	34.73
Adams et al. (1993)	18.50	25.11	37.21	95.06
Parks/Hardie (1992)	175.00	n.a.	n.a.	n.a.

Sources: Moulton and Richards, 1990, appendix, table 1A; Adams et al., 1993, table 1, p. 79; Parks and Hardie, 1992, figure 2, p. 40.

Finally, the extent to which large-scale tree planting can be a 'permanent solution' to the carbon problem is problematic, since there are presumably limits to the extent to which the earth can maintain forest stocks. Sedjo (1989) noted that it may be most useful to view tree planting as a measure to 'buy time' during which more permanent solutions to the basic fossil fuel nature of the carbon problem are developed. Richards (1993) has been especially sensitive to the intertemporal nature of the potential sequestration of carbon, pointing out that the expected life of carbon sinks may be limited and asking the question of how transitory sinks are to be valued, if at all. Although the methodology and empirical estimates have come a long way in the past five years, there are still issues that have not yet been addressed and/or completely resolved.

3 Carbon Sequestration through Forest Management

The ability of traditional forest management to influence the level of carbon sequestered is significant for some types of management and negligible for others. As demonstrated in the previous section, forest management which involves tree planting and rapid biological growth, can result in significant carbon sequestration. Similarly, activities which promote increased stocking, generate additional carbon sequestration. While fairly large amounts of information are available on the costs of planting and various forest management and silviculture practices[6] with the exception of tree planting there has been little work done on the cost of sequestering carbon using these various practices.

An exception is the work of Hoen and Solberg (1993) in which they assess the 'efficiency' of various strategies to sequester carbon in forest biomass in Norway. Their approach and use of units in the study makes direct comparison with the literature somewhat difficult, but the results seem to suggest relatively high costs per unit of carbon sequestered for most strategies examined. In a related piece of work, Hoen and Solberg find that while thinning is to be avoided, the carbon sequestration returns to forest fertilization generate marginal costs of about $71 per incremental tC captured for a Norwegian forest, even where current logging levels are to be maintained. Although this is a relatively high price, it suggests that additional work in this and other related areas may be warranted.

In the context of harvesting practices, the carbon implications and costs of carbon are beginning to be examined. For selective logging, practices that reduce logging damage to residual trees will generally be associated with reduced releases of carbon associated with harvests. For example, Putz and Pinard (1993) estimated that modest modifications in harvesting techniques in the selectively logged forests of Sabah via the introduction of reduced impact logging, could result in retaining 46 more tons of carbon per ha after 15 years at a cost of only about $3/tC. However, the applicability of this finding to the USA and the temperate forest in general is limited, in part because, while the typical tropical forest harvest regime involves selective logging, most temperate forest harvesting involves clear cutting.

Management techniques that maximize forest biomass growth appear, on the surface, to offer the greatest promise of carbon sequestration. Turner et al. (1993) summarize the findings of a number of studies as suggesting that the least promising silviculture practices, from a carbon storage point of view, are those such as thinning, fertilization and other stand improvement treatments. It is clear that many silvicultural practices are designed to improve some aspect of the wood and/or tree form and may do little to increase carbon sequestration. However, the longer term indirect effects may have significant carbon implications. For example, management techniques that allow greater portions of the forest to be harvested and converted to long-lived wood products may sequester a greater volume of carbon in the long term in both the forest itself and in the forest product stock. Hence, for example, although thinning for pulpwood may not increase the total volume on the stand, it may generate a greater total amount of sequestered carbon in the long run, if it allows

---

6 Dixon and Krankina, 1992; Sedjo, 1983.

more forest stands to produce solid wood products and a greater portion of the inputs of short lived paper products to be generated by thinnings and wood residues. However, the comparative carbon implications and costs of this hypothesis have not been examined in any depth.

4 Sustainable Forest Management Techniques: Including Long Rotations in Major US Forest Types

The economics of using sustainable-yield forest management techniques for sequestration of carbon have been developed in many of the earlier cited studies. The use of long rotations has not been examined in the context of the carbon issue, probably because even a cursory inquiry suggests that the costs are likely to be very high. It is clear that the substitute of long rotations for short ones will increase on average the volume of carbon sequestered. Thus, a regulated forest, *e.g.* one with an equal representation of each age class, will have a higher average biomass and hence higher carbon content, than would the same species with a shorter rotation. For example, the average forest age in a regulated forest with a 100 year rotation will be 50 years whereas in a regulated forest with a 50 year rotation, the average age will be only 25 years. Lower average age, of course, means lower timber and biomass volumes and less carbon sequestered.

However, as is well demonstrated in the literature, the financial costs of extending the rotation beyond the financially optimal are usually large, due to the presence of a positive discount rate. An example of the negative effects of long rotations on financial returns independent of the carbon question are discussed by Sedjo (1991). Only if the discount rate is zero can the rotation be extended. Thus, although it appears that long rotations involving larger volumes are desirable from a carbon sequestration perspective, this is unlikely to be a low-cost method to sequester carbon.

Less clear are the carbon implications of some of the proposed sustainable forest ecosystem management proposals. Obviously, if harvests are delayed, there will be more captive carbon remaining on the site. However, reduced harvests in some places usually result in increased harvests elsewhere.[7]

In the USA, the introduction of a modified selective harvesting regime, such as some of those being proposed for the 'new forestry' in the Pacific Northwest would, no doubt, reduce the carbon releases for a given area. However, reduced harvests in

---

[7] Sedjo, 1993.

one area only imply increased harvests elsewhere. Thus such an approach is unlikely to generate any net increase in carbon storage. Modified tropical forest harvest, by contrast, does not involve a reduction in the harvest per land area, but rather involves decreasing harvest damage to the residual forest.

Some elements of the issue of the comparative carbon sequestered in an old growth forest vs. that captured through harvests and long-lived wood products are developed in Harmon et al. (1990). They argue that old growth forests are superior to harvest and planting approaches in their ability to hold carbon. Kershaw et al. (1993) demonstrate that the ability of an old-growth forest to sequester carbon is dependent upon a host of circumstances. Furthermore, as developed in the Harmon et al. accounting framework, the carbon sequestered is dependent upon assumptions regarding the life of long-lived wood products, the amount of soil carbon loss at harvest and the amount and rate of decomposition of debris on the site following harvest. All of the above are variable or known only with a great deal of uncertainty.

5 Recycling Wood and Paper Products for Fossil Fuel Substitutes as well as for Recycled Paper Products

As discussed earlier, the forest sector carbon budget of the USA[8] uses a modelling approach which examines the implications on carbon sequestration of alternative recycling levels. Not surprisingly, the model finds that higher levels of recycling result in reduced demand on the forest stock and, if the forest stock is independent of the level of demand and price, increased amounts of captive carbon. The 'forest sector carbon budget' approach allows the forest stocks and flows to be systematically traced through time.

Increased recycling of wood and paper products offers the promise of increasing the amount of carbon sequestered in the forest by reducing demand for the forest stock, thus allowing a larger standing inventory, and thereby increasing the carbon captive in the forest.[9] Recycling would decrease the pressures to harvest for pulpwood, allowing the forest stands to mature. By increasing the average age and size of the inventory, the forest system would increase the carbon sequestered. If, however, investments in tree plantations are a function of demand for virgin wood fiber, the independence between the forest volume and its utilization is broken. In this situation the long-term volume cannot be viewed simply as the carbon saved by recycling,

---

8 Turner et al., 1993.

9 Adams and Haynes, 1991.

since investments in the growth of the forest will change. Thus, should recycling also reduce forest investments, all or part of the anticipated increase in forest stock might be offsetted.

6 Discussion and Recommendations

A major weakness of most of the attempts to examine the economics of managing carbon via forestry is that the focus is on the costs of undertaking a particular action, *e.g.* tree planting for purely carbon sequestration purposes. An alternative approach would be to ask the question of what additional costs would be required to generate additional carbon sequestering in the context of the tree planting that is already underway. For example, how much subsidy would be required to generate an additional tree planting that would result in the sequestration of (say) 100 mln. tons? The question of the incremental response of private sector tree growers to additional incentives for tree planting has not been investigated.

Nevertheless, this chapter demonstrates that there has been the development of considerable literature over the past several years examining the question of the economic potential for forests and forest management to offset some of the expected rise in the level of atmospheric carbon. In some topic areas the research is well advanced. In fact, in the examination of the economics of carbon sequestration by tree planting. Especially in the USA we can observe several 'generations' of research, each building upon earlier contributions. In other areas, by contrast, only very tentative first steps have been taken.

One of the most advanced areas is the development of cost curves for carbon sequestration through tree planting and with some forest management practices. Advances are being made both in estimating the rising cost functions and in the integration of some of these results into quite comprehensive and sophisticated models. In addition, the cost curve estimates appear to be converging. Also, the process of rationalizing the differences among the various estimated curves has provided insights suggesting the sources of the differences and thus likely areas for productive additional research. Important differences appear to be found: in estimates of carbon sequestered, especially in forest soils; in including and/or estimating the land costs; in estimates of the opportunity costs of the various lands taken from other uses and converted to forests; as well as in forest productivity and growth. Early work that suggested little potential for carbon sequestration through tree planting tended to ignore changes in soil carbon as well as to not recognize the rapid growth that could be achieved in young forests.

References

Adams, R.M., D.M. Adams, C.C. Chang, B.A. McCarl, and J.M. Callaway, 1993. *Sequestering carbon on agricultural land: A preliminary analysis of social cost and impacts on timber markets*, Contemporary Policy Issues, XI.

Adams, R.M., C.C. Chang, B.A. McCarl, and J.M. Callaway, 1992. 'The role of agriculture in climate change: A preliminary evaluation of emission-control strategies'. in: J.M. Reilly and M. Anderson (eds.), 'Economic Issues in Global Climate Change: Agriculture, Forestry and Natural Resources', Westview Press, Boulder.

Adams, D.M. and R.W. Haynes, 1991. 'Softwood timber supply and the future of the Southern forest economy', *S.J. of Applied Forestry*, 15, p. 35.

Andrasko, K.J., K. Heaton, and S. Winnett, 1991. *Evaluating the costs and efficiency of options to manage global forests: A cost curve approach*, Prepared for Technical Workshop to Explore Options for Global Forest Management, Office of the National Environment Board, Bangkok, Thailand, April 1991.

Binkley, C.S. and G.C. van Kooten, 1993. 'Integrating climate change and forests: economic and ecological assessments', *Climate Change.*

Cline, W. R., 1992. *The economics of global warming*, Institute for International Economics, Washington, DC.

Darmstadter, J. and A. Plantinga, 1991. *The economic cost of CO_2 mitigation: A review of estimates for selected world regions*, RFF Discussion Paper ENR 91-06, 60 pp.

Dixon, R.K. and O.N. Krankina, 1992. *Forestation and forestation management options and their costs at the site level: Selected case studies*, ERL, USEPA, Corvallis OR.

Dixon, R.K. and O.N. Krankina, 1993. 'Forest fires in Russia: Carbon dioxide emissions to the atmosphere', *Canadian Journal of Forest Research*, 23(4), pp. 700-5.

Dixon, R.K., K.J. Andrasko, F.A. Sussman, M.A. Livinson, M.C. Trexler, and T.S. Vinson, 1993. 'Forest sector carbon offset projects: Near-term opportunities to mitigate greenhouse gas emissions', *Water, Air, and Soil Pollution*, 70, pp. 561-77.

Dixon, R.K., J.K. Winjum, K.J. Andrasko, J.J. Lee, and P.E. Schroeder, 1993. 'Integrated systems: Assessment of promising agroforest and alternative land-use practices to enhance carbon conservation and sequestration', *Climate Change.*

Dudek, D.J. and A. LeBlanc, 1990. 'Offsetting new CO_2 emissions: A rational first greenhouse policy step', *Contemporary Policy Issues*, 8, pp. 29-42.

Faeth, P., C. Cort and R. Livernash, 1993. *Evaluating the carbon sequestration benefits of sustainable forest projects in developing countries*, World Resource Institute Report, Washington, DC.

Harmon, M.E., W.K. Farrell, and J.F. Franklin, 1990. 'Effects on carbon storage of conversion of old-growth forest to young forests', *Science*, 247, pp. 699-702.

Houghton, R.A., J. Unruh and P.A. LeFebure, 1991. 'Current land use in the tropics and its potential for sequestering carbon', in: D. Howlett and C. Sargent (eds.), 1991. *Proceeding for Technical Workshop to Explore Options for Global Forest Management*, IIED, London, UK, pp. 279-310.

Hoen, H.F. and B. Solberg, 1993. *Cost efficiency and optimal management strategies for sequestration of atmospheric carbon in forest biomass - Overview of analysis and modelling experiences from Norway*, Paper for the International Symposium on Systems Analysis and Management Decision in Forestry, Valdivia, Chile.

Kershaw Jr., J.A., C.D. Oliver and T.M. Hinckley, 1993. 'Effect of harvest of old growth douglas-fir stands and subsequent management on carbon dioxide levels in the atmosphere', *Journal of Sustainable Forestry*, 1, pp. 61-77.

Krankina, O.N. and R.K. Dixon, 1993. 'Forest management options to conserve and sequester terrestrial carbon in the Russian Federation', *World Resource Review*, 1993.

Marland, G., V. Dale, R. Graham, R. Luxmoore, S. Marland, S. McLaughlin, R. Norby, W.M. Post, T. Tschaplinski, J. Tuskan, and L. Wright, 1993. *Carbon dioxide offsets with forests and energy: Forest management for fixing and sequestering carbon*, Presented to the Second United States/Japan Workshop on Global Change, 1-3 February 1994, Honolulu, Hawaii.

Moulton, R. and K. Richards, 1990. *Costs of sequestering carbon through tree planting and forest management in the United States*, US Department of Agriculture, Forest Service General Technical Report WO-58. 17 pp.

Nordhaus, W.D., 1990. 'An intertemporal general-equilibruim model of economic growth and climate change', in: D.O. Wood and Y. Kaya (eds.), 1990. *Proceedings of the Workshop on Economic/Energy/Environmental Modelling for Climate Analysis*, 22-23 October 1990, MIT Center for Energy Policy Research, pp. 415-33.

Nordhaus, W.D, 1990. *To slow or not to slow: The economics of the greenhouse effect*, Unpublished manuscript, Yale University Economics Department, New Haven.

Parks, P.J. and I.W. Hardie, 1992. *Least-cost forest carbon reserves: Cost-effective subsidies to convert marginal agricultural land to forest*, Washington, DC.

Plantinga, A.J. and R. A. Birdsey, 1993. 'Carbon fluxes resulting from US private timberland management', *Climate Change*, 23, pp. 37-53.

Putz, F.E. and M.A. Pinard, 1993. 'Reducing the impacts of logging as a carbon-offset method', submitted to Conservation Biology.

Richards, K.M., 1992. 'Policy and research implications of recent carbon-sequestering analysis', in: J.M. Reilly and M. Anderson (eds.), 1992. *Economic Issues in Global Climate Change: Agriculture, Forestry and Natural Resources*, Westview Press, Boulder.

Richards, K.M., 1993. *Valuation of temporary and future greenhouse gas reductions*, presented at Western Economic Association Annual Meeting, Lake Tahoe NV.

Richards, K.R., R. Moulton and R.A. Birdsey, 1993. 'Costs of creating carbon sinks in the US' *Energy Conservation and Management.*

Richards, K.R., D.H. Rosenthal, J.A. Edmonds and M. Wise, 1993. *The carbon dioxide emissions game: Playing the net*, Unpublished paper.

Sampson, R.N., L.L. Wright, J.K. Winjum, J.D. Kinsman, J. Benneman, E. Kuersten and J.M.O. Scurlock, 1993. 'Biomass management and energy', *Water, Air and Soil Pollution*, 70, pp. 139-59.

Scurlock, J.M.O., D.O. Hall and J.I. Howe, 1993. 'Utilizing biomass crops as an energy source: A European perspective', *Water, Air and Soil Pollution*, 70, pp. 499-518.

Sedjo, R.A., 1983. 'The comparative economics of plantation forestry', Resources for the Future, 193 pp., Washington, DC.

Sedjo, R.A., 1989. 'Forests: A tool to moderate global warming?', *Environment*, 31 (1), pp. 14-20.

Sedjo, R.A., 1991. 'Forest resources: Resilient and serviceable', Chapter 4 in: K. Frederick and R.A. Sedjo (eds.), 1991. *America's Renewable Resources: Historical Trends and Current Challenges*, Resources for the Future, Washington, DC.

Sedjo, R.A., 1993. 'Global consequences of US environmental policies', *Journal of Forestry*, 9, pp. 19-21.

Sedjo, R.A. and A.M. Solomon, 1989. 'Climate and forests', in: N.J. Rosenberg, W.E. Easterling, P.R. Crosson, and J. Darmstadter (eds.), 1989. *Greenhouse Warming: Abatement and Adaptation*, RFF Proceedings, Washington, DC, Resources for the Future, pp. 105-20.

Swisher, J.N., 1991. 'Cost and performance of CO_2 storage in forestry projects', *Biomass and Energy*, 1(6).

Trexler, M.C., P.E. Faeth and J.M. Kramer, 1989. *Forestry as a response to global warming: An analysis of the Guatemala agroforestry and carbon sequestration project*, World Resource Institute, Washington, DC, 65 pp.

Turner, D.P., J.J. Lee, G.J. Koperper and J. R. Barker (eds.), 1993. *The forest sector carbon budget of the United States: Carbon pools and flux under alternative policy options*, US EPA, ERL, Corvallis/OR, 202 pp.

Winjum, J.K., R.K. Dixon and P.E. Schroeder, 1992. *An assessment of forest management practices for sequestration of carbon: Estimating global potential*, US EPA Report, ERL, Corvallis/OR.

Winjum, J.K. and D.K. Lewis, 1993. *Forest management and the economics of carbon storage: The nonfinancial component*, Research Paper, USEPA, Environmental Research Laboratory.

Wisniewski, J. and R.N. Sampson (eds.), 1993. 'Terrestrial biospheric carbon fluxes: Quantification of sinks and sources of CO_2', *Water, Air and Soil Pollution*, 70, 691 pp.

Wright, L.L. and E.E. Hughes, 1993. 'US carbon offset potential using biomass energy systems', *Water, Air and Soil Pollution*, 70, pp. 483-97.

23. Forests Absorbing Carbon dioxide Emission

One of the response strategies of the Dutch electricity generating companies towards global warming

HANS VERWEIJ*

1 Introduction

In response to the issue of global warming the Dutch government has committed itself to achieve a stabilization of CO_2 emissions at the 1989 levels by the year 1995 and a further reduction of 3 to 5% by the end of this century. Each sector in The Netherlands is expected to help in achieving this target.

The electricity generating sector is at present responsible for roughly 20% of the Dutch CO_2 emissions. Traditionally, the Dutch Electricity Generating Board (SEP) has taken a broad view of global warming and other environmental issues, *i.e.* the solutions should not be confined to the traditional domain of electricity supply only, but should be extended to the energy infrastructure in general and include 'no-regret' measures that compensate for actual CO_2 emissions. SEP believes that through a more integral approach higher energy savings and a greater emission reduction can be realized at lower overall costs. The initiative taken by SEP to provide funds for planting forests to assist the Dutch electricity generating companies in absorbing their actual CO_2 emissions will be discussed in this chapter.

2 Reducing CO_2 Emissions

Among the factors that contribute to the so-called enhanced greenhouse effect, energy use is without question the most important one. In the present situation in which the energy supply in The Netherlands is still largely based on the combustion of fossil fuels, three principal strategies can be employed to reduce the energy related CO_2 emissions:

- increasing end-use efficiency;
- changing the energy supply system; and
- limiting end-use by setting quotas.

* The author is associated with FACE Foundation, The Netherlands.

C. J. Jepma (ed.), The Feasibility of Joint Implementation, 329–333.

Although the implementation of measures to limit energy consumption still seem to be far away, they nevertheless may come on the agenda if the first two strategies do not yield the necessary reductions in the total volume of CO_2 emissions.

With respect to the first strategy, improving efficiency on the demand-side, SEP endorses activities of the electricity distributing companies to reduce electricity demand by increasing end-use efficiency. In The Netherlands, for example, energy is saved in industrial production by switching from a direct use of fossil fuels to the use of electricity. This effect can be strengthened if the use of fossil fuels at the source of electricity generation itself is also further reduced.

Because electricity generation in The Netherlands is already among the most efficient in the world, SEP cannot achieve great improvements in reducing CO_2 emissions in the short term. Furthermore, in the future, electrification will be an ongoing process, resulting in a 40% electricity growth within twenty years. In spite of this, SEP expects that national levels of CO_2 emissions from electricity generation and use can be stabilized at the 1990 level by the year 2010. In order to contribute to achieving this target, SEP concentrates on:

- improving power plant efficiency;
- industrial co-generation and district heating;
- alternative sources;
- load management;
- integrated resource planning; and
- carbon fixation.

3 The Storage of CO_2: the FACE Foundation

With respect to recovering, utilizing, storing and disposing CO_2, there are many techniques that can be applied. Several methods, like storage of CO_2 in old natural gas fields, have recently been investigated by SEP. As a result, from these investigations it was concluded that the required technology for this kind of conservation has not yet been proven applicable and will moreover be very expensive. At present, only nature can take care of CO_2 storage in an environmentally friendly way without having any impact on energy conversion efficiencies in technical installations. In this respect one can think of several 'natural' techniques of storing CO_2, of which trees especially are beneficial. So, carefully planted new forests, *i.e.* respecting nature and the socio-economic structures in the region concerned, will create significant positive side effects. So, even if the harmful effects predicted by some theoretical global warming models do not come about, tree planting can still result in positive effects.

To explore this no-regret option, in October 1990 SEP set up the Forests Absorbing Carbon dioxide Emission (FACE) Foundation, which aims to encourage the planting of new forests to sequester CO_2 from the atmosphere. The objective set by SEP is to create the absorbing capacity for a 600 MW modern coal-fired powerplant. Calculating a 25 year life-cycle of trees, the necessary area of forests was estimated at 150,000 ha.

Given the objectives of SEP, it is essential to take up as much CO_2 from the atmosphere as possible and to sequester it as long as possible. FACE has chosen three starting points:

- supplementarity;
- cost effectiveness; and
- sustainability.

Supplementarity

FACE best achieves its task if the afforestation would not have taken place without FACE funding. Only in such cases is there a demonstrable relation between the CO_2 emissions reduction, the sequestration in the new forests and the FACE funding.

Cost effectiveness

FACE partly funds the planting of forests and their maintenance for the first three years after the planting. FACE thereby pays a price for the CO_2-sequestration capacity of the planted forests achieved during its long growing period. The land, the forest and its products - including the timber - are in the possession of the forest owner. FACE considers the sequestration capacity as weil as the amount of CO_2 sequestered to be its exclusive concern.

The speed with which CO_2 is sequestered in a forest is a derived criterion associated with cost effectiveness. FACE stipulates that during a cycle of 25 years the planted forests must sequester a minimum of 5,500 Kg CO_2 per ha per year.

Sustainability

FACE is interested in preserving the new forests for as long as possible, which in legal terms is translated into a contract period of 99 years. Ecological and socio-economic preconditions must be satisfied in order to guarantee that a forest is preserved for such a long period. These preconditions also guarantee that the side effects of the new forest are as large as possible. These effects are, for example: nature conservation; watershed management; securing the supply of raw materials to a wood-processing factory; supplying fuel wood; and income generation for local people.

4 The Location of the FACE Projects

At present, The Netherlands has in total about 300,000 ha of forest. The possibilities to enlarge this area are very limited, which implies that for realising the target of planting 150,000 ha of forest, FACE has to look for other countries with a great potential for sequestration and positive side effects. These countries can be found in Central Europe, where forests have been destroyed by acid rain caused by industrialization, and in the tropics, where large areas of rainforests are heavily logged without rehabilitation. Therefore an appreciable part of the new forests, 125,000 ha, will be planted in tropical regions and about 20,000 ha in Central Europe. The remaining 5,000 ha to be afforested are located in The Netherlands. At present, FACE has already signed contracts for the 150,000 ha needed.

Table 23.1. CO_2-sequestration capacity of FACE-projects.

	Rotation Period (years)	CO_2/ha/year	Total area (ha)	Total sequestration (CO_2)
The Netherlands	100	650	5,000	3,250,000
Czech Republic	20	708	16,000	11,328,000
Malaysia	70	918	25,000	22,950,000
Ecuador	25	473	75,000	35,475,000
Uganda[a]	70	759	35,000	26,565,000
Total			156,000	99,568,000

a The project in Uganda is in preparation.

The table shows that the CO_2-sequestration capacity of rehabilitated rainforest is much larger than in other regions. The European forests become more attractive for CO_2 sequestration because of their longer rotation period, while the relatively short rotation periods of the forests in Ecuador give a lower result. The total amount of sequestration has been determined at about 100,000,000 CO_2 which amounts to the CO_2 emission of a 750 MW Dutch coal-fired power plant during 25 years (being the amortization period of a coal-fired plant), and which is about 130% of the target of FACE. It compensates for about 10% of the total CO_2 emissions of Dutch power plants.

An example of a rehabilitation project can be found in the FACE project in Sabah-Malaysia (25,000 ha, see Table 21.1), which supports the replanting of logged over rainforest in collaboration with the concession holder. First of all a contract has been signed with the State government of Sabah, which declares that the latter is aware of the objectives of FACE and that the project fits within the national forestry policy of Sabah-Malaysia. The contract partner reafforests 1,000 ha of logged-over forest every year during a period of 25 years. FACE also supports research to optimize the propagation of the young trees and the sustainability of the forest management. In the contract FACE only claims the CO_2-sequestration capacity; all the other benefits are good for the project partner, whose interest is to have a sustainable flow of timber from its concession over a long period.

5 Conclusion

The approach of SEP and FACE can be a basis for the elaboration of criteria for the acceptation of forestation programmes within the FCCC. Of course the forestry projects should fulfil certain criteria. The forests have to be established:

- in cooperation with the local and national government authorities;
- in cooperation with the local people; and
- within sustainable ecological conditions.

This cooperation should be on the basis of equality. It is important that the new forests first contribute to solving local problems and that they favour local objectives. When this condition has been fulfilled, forestation can be a part of the solution of the problem of a CO_2 producer.

With respect to JI, FACE recently considered the possibility to create an extension with another project in Malaysia. However, the Federal Government of Malaysia is not willing to give its agreement to such an extension as it wants to await international rules with respect to JI.

24. Why Joint Implementation Can Boost Demand Side Management in Developing Countries

WIM WILMS[*]

1 Introduction

In this chapter Joint Implementation will be discussed from a business perspective. In particular, the focus will be on the potential role JI can play in further developing Demand Side Management (DSM) in developing countries. Philips Lighting is very much involved in the development and production of the energy saving Compact Fluorescent Light bulbs (CFLs). These CFLs are good examples of products that can be used in DSM to improve energy efficiency in industrialized as well as in developing countries. Philips Lighting has already gained significant experience in replacing CFLs for incandescent lamps, so that knowledge about the product and the customer already exists.

However, in introducing CFLs the producing companies have thus far met significant barriers at the level of the customer as well as at the level of the national governments and the utilities: the product is new and still needs to gain credibility, whereas the purchase and replacement of CFLs also requires a whole new decision making process. In order to stimulate the introduction of CFLs in the market, large public CFL campaigns can provide the necessary assistance, as can be inferred from the evidence in industrialized countries which have already carried out such campaigns. According to the business community, JI can play a very important role in this respect.

From a business perspective, JI can, in stimulating DSM, be approached in two ways. Firstly, there is the perspective of the companies which produce the products suitable for DSM, like, for example, the producers of CFLs. Secondly, JI can be viewed from a broader perspective in which three point are relevant:

- the product;
- the customer who can buy the product; and
- the incentives for the customer to buy the product, *i.e.* the question whether the customer is convinced that the product will be beneficial for him.

---

[*] The author is associated with Philips Lighting BV, Product Management CFL, Support Team Energy Projects, Roosendaal, The Netherlands.

C. J. Jepma (ed.), The Feasibility of Joint Implementation, 335–343.

In this chapter, this second business perspective will be discussed. In other words, the product is DSM via CFLs; the customers are the consumers from all over the world, and our aim is to publicly improve the consumers' knowledge of the product, so that they will learn why this particular type of DSM might be beneficial for them.

In section 2 a very brief introduction to DSM will be given. Section 3 deals with the CFLs and tries to answer the question why CFLs have not yet been purchased by consumers on a large scale. Section 4 answers the question why not all electricity utility companies have implemented DSM yet. Section 5, finally, tries to point out what role JI can play to globally stimulating DSM, or, in other words, how JI can keep DSM going.

2 DSM and CFLs

In many studies on reducing emissions of CO_2 through DSM the replacing of CFLs for incandescent light bulbs is often recommended as a cost-effective way. Why can CFLs play such an important role in DSM? First, it can be pointed out that the development of CFLs has been thorough, which has resulted in several different types of bulbs and which has also improved the technical quality of the product. Customers can more easily choose between light bulbs with different power, size, *etc.*, whereas the uncertainty about the quality of the bulbs has become less. Second, although one disadvantage of CFLs which is often mentioned is the high price of the bulbs, comparing the replacement of CFLs for incandescent lamps as an energy saving measure is still much cheaper than, for instance, replacing an energy saving washing machine for an old one. Third, in contrast with incandescent lamps, CFLs have a long lifetime, approximately eight years. Given the fact that during one year incandescent lamps need to be replaced a few times, it is clear that, despite their relatively high purchasing cost, CFLs might in the end turn out to be considerably cheaper than incandescent lamps. Finally, CFLs simply result in high gains in energy efficiency: the lower amount of energy needed in households when using CFLs instead of incandescent bulbs roughly amounts to 75%. For these reasons, where in the past incandescent bulbs were the symbol for lighting, CFLs have now become the symbol of energy efficiency.

3 Why don't all consumers use CFLs?

Turning back to the reasons for not using CFLs, consumers might decide to purchase

CFLs because of: the lower electricity bills that result when using CFLs instead of incandescent lamps; the longer lifetime of CFLs; and, perhaps even more importantly, people might realize that by using CFLs they can really contribute to help saving the environment. Why then does not everybody use CFLs?

Basically, it seems plausible that consumers do not consider electricity prices and their electricity bills as key elements in their decision-making process with respect to buying lightbulbs. Generally, it seems to be very difficult to change the relatively rigid 'lamp purchasing behaviour' of simply buying incandescent bulbs in the nearest supermarket. People who are used to buying lamps in this way will not quickly move to more expensive ones, even if this might in the long term be financially attractive for them.

To illustrate this, Figure 24.1 shows the relationship between the electricity prices for homes (in ECU/100 kWh) in nine European countries in 1992 on the one hand, and the number of CFLs sold in the countries in 1992 on the other hand. Intuitively, one would expect to find a positive correlation between electricity prices and the amounts of the CFLs sold in a country. As can be inferred from the figure, however, such a correlation does not exist. In general, the use of CFLs is relatively low, except in The Netherlands and Denmark. Another striking result is that in The Netherlands, where electricity prices are relatively low (about ECU9/kWh), the amount of CFLs sold is relatively high, whereas in countries with high electricity prices, like Spain and Portugal, the amount of CFLs are the lowest of the countries investigated.

The reasons why consumers do not use CFLs in an amount as high as might be expected are basically: the present lack of credibility of the cost-saving aspect of CFLs (do the bulbs really save money?); the novelty character of the product (a CFL is a 'strange thing' compared to incandescent bulbs, which people have been using for nearly the last 100 years); and the new decision-making process consumers have to go through when purchasing CFLs. With respect to the latter one can, for instance, think of the difference between buying incandescent lamps in the supermarket and 'investing' in CFLs.

A prominent problem in this respect is the lack of information about the use of CFLs available in the households. The latter problem can be tackled via so-called Utility CFL campaigns, which are DSM programmes aiming at increasing the penetration of CFLs in homes via extensive information campaigns. Utility CFL campaigns should be set up by electricity utility companies in close cooperation with lighting manufacturers who assist them in promoting the product and in providing information to the consumers. This can, for example, be done with the help of financial incentives (the electricity bill), through providing information on television, in

Figure 24.1. The relation between the number of CFLs sold per home and the electricity costs for homes in 1992.

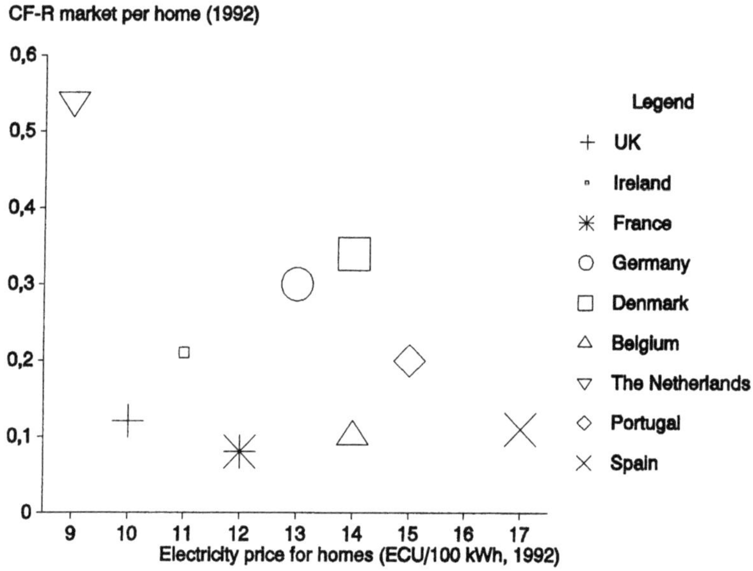

newspapers, *etc.*, and through the distribution of additional information by the national and local governments (for example, by providing calculations that show the cost-saving potential of CFLs). In the USA and in some European countries such campaigns have already started. In the next chapter, which deals with a typical DSM project in a developing country, the ILUMEX project in Mexico, it will be demonstrated how such a campaign can be carried out there.

In Figure 24.2 the response results of Utility CFL campaigns are illustrated. On the vertical axis some EU countries and Switzerland are indicated, and on the horizontal axis the number of CFLs installed in homes in 1992. It can be seen that in the countries where CFL campaigns have already actively been carried out, The Netherlands and Denmark, the effects on the amount of CFLs sold were enormous. The conclusion can be that, through Utility CFL campaigns, electricity utility companies, CFL producers and governments can indeed significantly increase the credibility of CFL: it can be demonstrated to the consumers that CFLs really save energy and money whilst it can furthermore be fairly easily shown that consumers can really help to improve the environment.

Figure 24.2. The average number of CFLs installed per home in some European countries upto and including 1992.

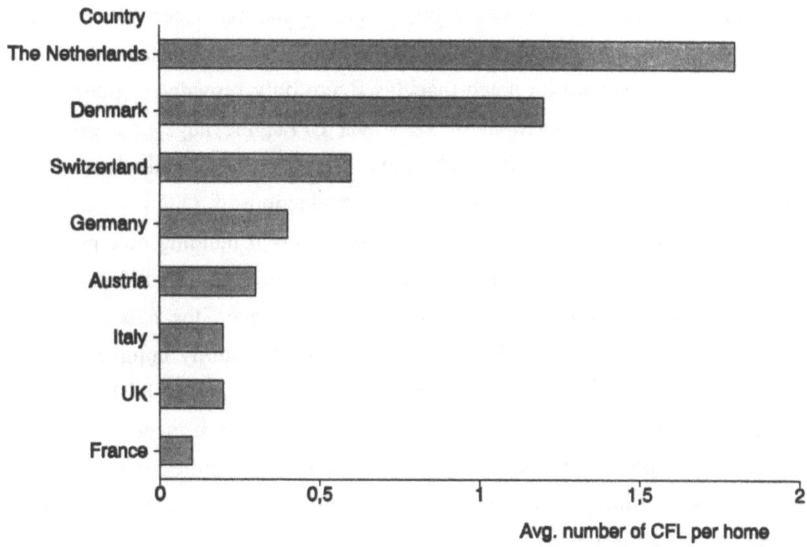

4 Why don't all Utilities Implement DSM?

So far we have focused on the question why CFLs are not yet being used by most consumers. One of the main conclusions of the former section was that the credibility of CFLs can be improved through DSM via Utility CFL campaigns. In such campaigns an important role should be played by electricity utility companies. Thus far, however, only relatively few electricity utilities have applied Utility CFL campaigns. At first sight this might seem to be a strange situation, since, as the following figures will show, the cost saving opportunity is relatively high.

Based on figures from the World Bank and the World Energy Conference,[1] the cost saving opportunity can be illustrated as follows. If, in order to add extra capacity to electricity production, an electricity utility company decides to build a new power plant, the costs (of generating and transmission of the additionally produced electricity) will amount to roughly between \$1,500 to \$4,000/kW in developing countries and between \$1,000 and \$2,000/kW in industrialized countries. If one furthermore

[1] IAEEL, 1993, p. 12.

assumes that the lifetime of the new power plant is roughly 20 to 30 years, and if the need for installation of this extra capacity would disappear through the replacement of CFLs for incandescent lamps - according to the expected lifetime, the CFLs need to be replaced approximately every eight years - the costs per kW (the costs of replacing CFLs for incandescent lamps) would only amount to about $500 (present value). Although it should be noted that this figure only provides a rough illustration of the cost-effectiveness potential of DSM via CFLs, the huge cost difference between the two options - creating additional capacity for effectively producing or reducing the need for this by large-scale introduction of CFLs - may be clear: investments in CFLs are at least 50% lower than those in building new power plants.

Looking at all this, it seems strange why DSM is currently not being implemented by electricity utility companies on a larger scale. However, the reasons why consumers do not use CFLs in their households also hold for utility companies. The first point again is very important: the credibility issue. Theories on DSM and its results are potentially very attractive, but does the option really work in practice? There has not been much proof yet. Secondly, for an electricity utility company that has exclusively been selling kW hours for its entire existence it may be a strange phenomenon to inform consumers about replacing their old light bulbs for CFLs: most utilities are not used to that and might need to change their strategies. Thirdly, and especially relevant for developing countries, DSM requires a completely new decision-making process. Except for the electricity utility company, there are many other institutions involved in the DSM programme, such as, for instance: the producer of the DSM product; the distributor of the product; the government of the host country (usually the department of energy); the energy regulation board; regional development banks; international institutions like the World Bank; and several study groups. All of these institutions may want to have some influence, which, of course, very much complicates the decision-making process. Moreover, it sometimes seems that the more studies are being carried out, the less projects are being undertaken. As a final complication, there might be a good chance that the institutions may decide to wait for each other's first move, resulting in a status quo in which no activities are being started. Because of these difficulties and barriers DSM is still not yet widely accepted.

If we look at the differences between industrialized and developing countries with respect to large DSM campaigns with CFL that have already been carried out, one can see on the one hand that some industrialized countries (Australia, Canada, Denmark, Greece, Ireland, Sweden, Switzerland, The Netherlands, UK and USA) have up until now (mid-1994) been very active; however, on the other hand, of the developing countries only Guadeloupe and Martinique have carried out, large DSM campaigns with CFLs. One may note, as the next chapter will show, that Mexico is also working on a DSM campaign, but that its project has not been carried out yet.

5 Why JI can boost Demand Side Management

In industrialized countries there are already some utilities which have quite a lot of know-how and experience with DSM campaigns with respect to CFL. The question is in how far JI can play a role vis-à-vis such types of DSM. One can think of the following possibilities.

First, an electricity utility company in an industrialized country can, via a DSM programme, cooperate with a utility company in a developing country, which, for instance, has a problem in generating a sufficient amount of electricity to meet its domestic electricity demand. Such a (JI) project can be carried out in a customer friendly way (*i.e.* beneficial to the developing country). The industrialized country's company might be credited some way or another, or should at least be given an incentive to invest.

Second, via JI DSM can be applied through *real* projects. So far, many studies on DSM have been carried out, but it now becomes important that parties 'start walking instead of talking' and make themselves confident and experienced with the concept. Only by implementing real projects will developing countries find out that DSM actually works: the peak load decreases!

Third, through JI the credibility of DSM (products) can be increased. The existing doubts in developing countries with respect to DSM can be lowered when, via JI DSM projects, the experience and know-how of utility companies and possibly governments in industrialized countries can be positively transferred to the utility companies in developing countries.

6 Conclusion

Summarizing, DSM can significantly improve the world's energy saving potential. However, due to a lack of credibility, scepticism, doubts, complicated decision-making processes, *etc.*, the implementation of DSM has thus far mostly been limited to industrialized countries. It has been shown that DSM can benefit significantly from large-scale Utility CFL campaigns being carried out by electricity utility companies in cooperation with national governments. Such campaigns have almost exclusively been carried out in industrialized countries. Further stimulating Utility

CFL campaigns and providing electricity utility companies in developing countries with information and experience seem to be a prerequisite for a successful implementation of this type of DSM. The role of JI can in this respect be very important, because through this concept industrialized countries' experience can fairly easily and cost-effectively be transferred to developing countries. The option is furthermore beneficial to the consumers in developing countries as well as to the country itself.

Reference

IAEEL, 1993. 'Light for Developing Nations', *International Association for Energy Efficient Lighting Newsletter*, issue no. 3, vol. 2, pp. 11-2.

25. A High Efficiency Lighting Project

The Mexican ILUMEX Project

LUIS LUZURIAGA*

1 Introduction

This chapter will discuss one of the projects mentioned by Roland and Haugland in Chapter 26: the Mexican ILUMEX project. This project has been set up by a cooperative action of the Mexican government, the Norwegian government and the Global Environment Facility (GEF), and aims to replace 1.7 mln. incandescent light bulbs with energy saving fluorescent light bulbs (these include compact and circular fluorescent light bulbs, in this chapter indicated by CFL). The project can provide valuable experience for the setting up of future JI projects, as it is currently recognized as a JI simulation project: *i.e.* a project that has already been initiated and which can, because of the similarities with JI, be viewed from a JI perspective.

The several aspects of the implementation of the ILUMEX project will be dealt with in this chapter. Section 2 analyzes the country and sector background of the project; section 3 deals with the objectives and description of the project; section 4 discusses the cost of the project and the available funding to finance implementation; section 5 deals with the institutional aspects of the project implementation; sections 6, 7 and 8 shortly discuss the procurement and disbursement of the funds, the project sustainability, and the lessons to be learned from previous GEF projects, respectively; section 9, finally, provides a survey of the already agreed actions, the environmental aspects and the potential benefits and risks of the ILUMEX project.

2 Country and Sector Background

In 1991 Mexico's GDP growth was 3.6% and in 1992 2.6%, while its per capita GDP was $3,030 and $3,470 in 1991 and 1992, respectively. Since per capita GDP was $2,990 in 1989, Mexico is eligible for GEF financing.

The proposed ILUMEX project underscores Mexico's commitment to environmental protection. The country has a satisfactory national environmental policy

* This chapter is based on a document of the World Bank/GEF, 1994. Report No. 12448-ME, composed by Luis Luzuriaga (Principal Power Engineer, Task Manager), Richard Clifford (Financial Analyst), Rudolf Van Puymbroeck (Legal counsel), Luis Cosenza (Power Engineer, Consultant) and included John Skjelvic and Rolf Selrod, representing the government of the Kingdom of Norway.

C. J. Jepma (ed.), The Feasibility of Joint Implementation, 345–357.

framework, and with the assistance of the World Bank it prepared an environmental action plan,[1] which has been completed in mid-1994. In particular, and also with assistance from the Bank, Mexico has begun to take firm steps to reduce contamination from vehicles, including the introduction of unleaded gasoline, mandatory use of catalytic converters for all new vehicles, and strict vehicle emission inspections twice a year. Mexico is a signatory of the FCCC and its government is committed to formulate and implement national and regional programmes to mitigate climate change. Under the Montreal Protocol, Mexico has decided to phase out ozone depleting substances, and agreed to meet timetables for industrialized countries; this initiative is being supported through two projects financed with Montreal Protocol funds and for which the Bank is the implementing agency. More importantly perhaps, public awareness of environmental problems has been growing and the government's concern for the environment is widely supported by the general public.

In May of 1992, SEDUE (the ex-Ministry of Urban Development and Ecology) was absorbed within the newly created Secretaría de Desarrollo Social, SEDESOL (Ministry of Social Development). This made it possible, among other things, to have a better coordination of the government's environmental policy. Within SEDESOL, the development and implementation of Mexico's environmental policy is the responsibility of two decentralized agencies: the Instituto Nacional de Ecologia (the National Environmental Institute, INE); and Procuraduría de Protección Ambiental (the Environmental Protection Office). The World Bank is assisting Mexico in strengthening its environmental institutional capacity through the Mexico Environmental project, which includes a \$25 mln. biodiversity component financed by the Global Environment Trust fund (GET).

Demand for electric energy in Mexico is high: it was 100 TWh in 1992 and is projected to grow at 5.3% per annum, resulting therefore in the need to add 14,000 MW over the next ten years. Massive investments - in the order of \$3 bln. per year for generation, transmission and distribution - will be needed to meet this growing demand; part of these investments are expected to be made by the private sector, as the Electricity Law, amended in 1992, now permits independent power generation. Since Mexico's installed generating capacity is about 80% thermal, there is concern over the gaseous emissions from thermo-electric power plants and their deleterious effect on the environment. Current estimated CO_2 emissions from thermal power stations amount to 57 mln. tons per year; similarly emissions of sulphur dioxide (SO_2) and nitrogen oxide (NO_x) amount to 1,070,000 tons per annum and 115,340

[1] A detailed description and assessment of Mexico's environmental issues can be found in the World Bank, 1992.

tons per annum, respectively. The federally owned electric power utility (CFE) has implemented, with World Bank assistance, supply-side management and demand-side management (DSM) actions. It is also expected that the newly created Energy Regulatory Agency will establish policies to further enhance DSM activities and require Integrated Resource Planning as opposed to Least Cost Planning now practised by CFE.

Electricity tariff increases over the last three years have resulted in an average price for 1993 of ¢6.92/kWh, or about 83% of long-run marginal costs (LRMC). The average price for residential consumers was ¢6.65/kWh, with a range that varied between ¢5.3 and ¢15.3 per kWh, involving a significant cross subsidy from residential consumers, with medium to large consumption, to other consumers. This situation is being corrected, as CFE is committed to gradually eliminate by 1997 the existing large cross subsidies among different types of consumers and to raise its average price to reach LRMC by the same year.

In August 1990, CFE established a trust fund (FIDE) with funds provided by power utilities, labour unions, contractors and suppliers to support and finance programmes related to energy conservation in the power sector. To date, FIDE has managed funds totalling approximately $16 mln., which have been used to finance: pilot and demonstration projects; dissemination of their results; and training in energy conservation. In addition to partially funding FIDE, CFE has directly promoted and financed several energy conservation projects, such as: the use of more efficient light bulbs (sodium vapour) in the public lighting system; the financing, through a trust fund, of the thermal insulation of homes in areas with hot weather to reduce the air conditioning load; assistance to the private sector to identify measures to improve energy efficiency; and seven small pilot projects to replace incandescent bulbs with CFLs in Hermosillo (two projects), Puebla, Querétaro, Valladolid, Chetumal and Mexico City.

3 The ILUMEX Project

3.1 The project objectives

The objectives of the ILUMEX project basically are to:

- demonstrate the technical and financial feasibility of reducing GHG emissions and simultaneously reduce local environmental contamination through the widespread installation of high efficiency lighting;

- build the institutional capacity for technological change and energy conservation;
- provide a replicable model for DSM in Mexico and elsewhere in the developing world; and
- strengthen CFE's capacity to practice DSM on a sustainable basis.

Although FCCC Parties may implement policies and measures to reduce emissions of GHGs jointly with other Parties, the criteria for JI are, however, yet to be developed and decided upon by the Parties. To assist in the development of possible criteria, the Norwegian government (see also Roland and Haugland, chapter 26) has decided to co-finance the ILUMEX project, which might serve as a demonstration vehicle for elements of a JI scheme. The agreement of the governments of Mexico and Norway on the co-financing of this project in no way implies the positions that both governments might take in relation to JI under the FCCC.

3.2 Project description

The project will replace approximately 1.7 mln. incandescent bulbs with CFLs in the cities of Guadalajara and Monterrey. CFLs require only about 25% of the energy needed by incandescent bulbs to produce the same lighting level and last up to 13 times longer (up to 10,000 hours). Guadalajara and Monterrey are Mexico's second and third largest cities with approximately 550,000 residential electricity consumers each; they also have different climates and economic structures and should, therefore, provide valuable information for extending the programme to other areas. CFE is responsible for the distribution of electricity in both cities, where there are customer service centres that are in daily contact with the consumers, a necessary condition for the successful implementation of the proposed project.

The project includes:

- the acquisition of CFLs (77% of the project costs, including contingencies);
- the purchase of vehicles, metering equipment, personal computers, office equipment and sales stands (2%);
- consultant services for canvassing, marketing, testing, auditing and project evaluation (4%);
- the cost of engineering, monitoring and evaluation, billing and accounting (8%); and

- the direct costs of implementing the project, including fully dedicated staff in both cities, and office and administrative expenses (9%).

There are no policy or institutional reforms needed to ensure an effective project implementation. However, experience gained from the small pilot projects mentioned in section 2 points to the need for baseline assessments and a comprehensive evaluation procedure. For the proposed project, a baseline assessment will be carried out through specialized consultants. Their work should provide the basis for a suitable evaluation process, particularly with respect to induced sales that result from project related activities (the 'free rider effect', see Roland and Haugland, chapter 26).

The project incorporates the following features:

- the benefits will be shared with the participants, as is common in many other countries, through a rebate of approximately 63% (on average) of the total cost of the CFLs (which includes overhead costs);
- participants will pay for CFLs either in cash or under a deferred payment plan of 24 months as part of their electricity bill, paying an interest rate equivalent to that prevailing for government treasury bills (CETES) (at the end of the present project, the balance of reflows of funds would be approximately $9 mln.);
- technical standards and specifications have been developed to ensure that the CFLs will operate properly and reach their projected lifetime despite a power system that is subject to significant voltage variations; and
- acceptable CFL performance guarantees will be provided for both the customer and CFE.

4 The Funding of the Project

4.1 Rationale for GEF funding

The GEF classifies projects into three types. Type 1 include projects that are economically viable on the basis of domestic benefits for and costs to the country itself. These projects are normally not eligible for GEF financing, unless the GEF Participants agree that a compelling case has been made that, despite attractive rates of return, the operation would not proceed without GEF involvement. The proposed project has attractive rates of return, and it could be expected that it would be implemented purely for its cost-effectiveness, and that the use of the CFLs would spread, even in the absence of the project.

However, experience confirms that unless consumers are shown how they can benefit from CFL use, the process of change is extremely slow. A major problem is the high initial investment, since CFLs typically cost at least ten times as much as the incandescent bulbs that they replace. Furthermore, there is a need to inform, and more importantly, convince the consumers that they will in fact eventually reap the expected benefits. CFE's own small pilot projects have demonstrated that, given a suitably designed project, consumers will accept the use of CFLs. However, these projects have been too small to have an impact on the nation-wide use of CFLs. GEF participation is, therefore, essential to demonstrate that benefits to the global environment and to Mexico can be realized.

Based on the above considerations, the proposed project has been designated as a Type 1 project. The project has been considered and endorsed by the GEF Scientific and Technical Advisory Panel (STAP), and has been authorized as part of a work programme approved by the GEF Participants at their December 1991 meeting.

4.2 Project costs and financing

The total project costs (including price and physical contingencies) have been estimated at $23 mln., of which:

- $17.6 mln. for the purchase of the CFLs;
- $0.4 mln. for the purchase of vehicles, metering equipment, computers, office equipment and sale stands;
- $1.0 mln. for consultant services (interviewing, marketing, monitoring and evaluating, laboratory testing and auditing);
- $1.9 mln. for engineering and project support; and
- $2.0 mln. for direct project implementation.

The project will be financed with a grant from the GET of $10 mln., a grant from Norway of $3 mln., and of $10 mln. from CFE's own resources. The grant funds will be used to partly finance the purchase of CFLs and will subsequently be transferred to the participants of the programme as rebates for the purchase of CFLs.

5 Project Implementation

For the design of the proposed project, CFE has relied on the results of market studies made in Guadalajara and Monterrey, and on the experience gathered on residential lighting in the pilot energy conservation projects, as well as in the thermal insulation improvement projects. These projects have illustrated the advantages of proceeding through Implementing Units, financed through trust accounts ('Fideicomisos') set up with the Mexican National Bank of Public Works and Services (BANOBRAS). This has been shown to be an efficient implementation process, and has allowed CFE to retain control of the projects and to introduce timely adjustments to ensure their success.

This process also guarantees that the funds, including those resulting from the credit sales to the participants, are kept separate from those that enter into CFE's own treasury and that they are only used for the purposes of the project. The project will therefore be implemented through two Implementing Units, one for each city, that will administer trust accounts opened in BANOBRAS. Each Implementing Unit will be directed by a Technical Committee with five members, of which four are CFE's staff members. The Technical Committee will appoint for each Unit a manager who will be responsible for the day-to-day management of the project in the city; the manager, and the personnel under his responsibility, will be hired by, and paid with funds from the trust account. CFE will also establish a project support and monitoring group to provide guidance to the project and to provide day-to-day support and supervision through a Project Coordination Unit (PCU). A high level Project Advisory Committee will oversee the execution of the project.

CFLs will be sold in the sale stands that CFE will install in every customer center in the two cities where consumers pay their electricity bills and in selected shopping centres, and as required, by using door-to-door sales. Collection of payments for CFLs sold on terms will be done using CFE's customer service facilities and will be included in the customer's electricity bill. CFE's systems and experience on these matters are more than adequate to ensure that there will be no problems in the areas of billing and accounting.[2]

CFE has proposed to restrict the project to the residential sector since the other sectors, which include industrial and commercial users, do not require special incentives to make CFLs attractive; in addition, CFE does not wish to be seen by the

[2] For a more detailed description of the project organization, operating guidelines, project funding, project monitoring and evaluation, etc., the reader is referred to: World Bank/GEF, 1994.

commercial distributors of CFLs in Mexico as their competitors. In order to promote the use of CFLs more widely, the project will fund a marketing campaign to foster use of CFLs, sold through normal retail outlets, among the commercial and industrial customers.

CFE further plans to direct the project to the low consumption users in the residential sector to provide a relief, by means of more efficient use of electricity for lighting, when future tariff increases are implemented. This should not reduce the energy savings produced by the project: experience shows that through low consumption users have fewer light bulbs in their homes, they make equal or more intensive use of them compared to high consumption users because low consumption users spend more time in their limited number of rooms lit. At present, electricity sales to these customers result in a net loss for each kWh, and CFE understandably wishes to reduce its losses. Nonetheless, since energy rates for low consumer users are heavily subsidized, these customers would normally be the least inclined to participate in the project. Consequently, sales will be monitored closely to ensure that they do not proceed painstakingly slowly while waiting for these low consumption customers to participate. Targets have therefore been developed and failure to reach them will trigger off more aggressive marketing strategies.

It is estimated that all the CFLs will be purchased and installed in two and a half years after the project start (at the time of writing the project was expected to start in mid 1994), that the evaluation of the operation will be completed six months later, and that a final project survey among the participants will be carried out a year after completion. Consequently, the duration of the project has been estimated to be three and a half years: the project completion date is estimated to be 31 December 1997. However, as some of the CFLs will be sold on deferred payment terms, recovery of funds will require up to four and a half years. BANOBRAS will be responsible for the management of the grant funds, for keeping project accounts, for arranging audits by independent auditors, and for sending annual audit reports to the Bank.

6 Procurement and Disbursement

In order to profit from economies of scale, CFLs will be purchased using both GET and Norwegian grant funds, in no more than three bid packages of not less than $4 mln. each, under International Competitive Bidding (ICB) in accordance with World Bank guidelines for procurement and using World Bank standard bidding documents. Procurement documents will be subject to prior review by the World Bank. The first contract was planned to be awarded early in 1994, and the second and third before the stock of CFLs is depleted.

It is expected that there will be keen competition among manufacturers, including local companies that at present produce about five mln. CFLs annually, mainly for the export market. Purchase of vehicles, computers, metering equipment, sales stands and office equipment will be financed with funds provided by CFE and will follow local procedures. The hiring of consultants will also follow local procedures as they will be financed with CFE's contribution. In view of the limited number of expected contracts to be financed by the grants, disbursements will be by direct World Bank payment to suppliers. Disbursement of both grants will be made on a *pari passu* basis, and would be completed in three years; the closing date of the grants will be 31 December 1996. Administration and disbursement of the grant from Norway will be carried out by the World Bank together with the GET grant.

7 Project Sustainability

The project could be replicated using the reflow of funds from the sale of the CFLs, provided that CFE would continue sharing its savings with the participants and would also contribute fresh funds to the subsequent projects. CFE has agreed to expand the scope of the project, if required, until all funds from both grants have been transferred to the participants in the form of rebates for the purchase of CFLs. CFE furthermore agreed that if the project is successful, the Implementing Units will continue their activities beyond the scope of the present project, until the balance of the reflow on funds of about $9 mln. is completely exhausted.

8 Lessons from Previous GEF Projects and Previous Bank Involvement

CFE's experience from its previous small pilot projects indicates that:

- it is necessary to share a significant part of the benefits of the project with the users through a rebate as the cost of the CFLs is otherwise perceived as too high by potential participants;
- it is important to have a well defined marketing and promotional strategy;
- certain types of CFLs are preferred, particularly the 22 watt circular lamp; and

- to penetrate the low consumption residential market, door-to-door sales are most effective.

Furthermore, the World Bank's experience in the power sector indicates that: agreement should be reached early on draft ICB documents and the Ministry of Finance and Public Credit must be committed to the project in order to allow CFE to include it in its investment programme. Agreement on bid documents for ICB was reached, the Ministry's commitment has been secured, and the other mentioned lessons have been incorporated in the project design.

9 Actions, Effectiveness, Environmental Aspects, Risks and Benefits

9.1 Agreed actions

With respect to the ILUMEX project, agreements were reached on:

- The final text of the Operating Guidelines for the project, which will constitute the norm for the execution of the project. Among other things, the Guidelines: provide guidance to CFE, BANOBRAS and the Implementing Units (IUs) for the execution of the project; establish the percentage of the rebates and provide flexibility for adjusting them and for shifting emphasis among the various categories of participants, considering the results of the periodic reviews of the project; permit the funds to be distributed differently than the original plan if the project in one of the two cities is seriously lagging; and state the periodic goals to be achieved, and the monitoring and evaluation plan. Any modifications to the Operating Guidelines have to be satisfactory to the World Bank.
- The text of the document that will establish the trust deeds for the Implementing Units in Guadalajara and Monterrey, which include their obligation to observe the Operating Guidelines.
- The obligation of holding together with the Bank annual reviews and a mid-term review, and the targets to be achieved; at the time of the mid-term review, the Bank will have the option of cancelling the balance of both grants if sales have not reached at least 70% of the agreed total targets.
- The commitment of CFE of preparing semi-annual progress reports for the project, a project completion report, and of disseminating the results of the project through a seminar directed to Mexican and international audiences.

- The obligation of CFE to expand the geographic area of the project, if needed, to sell more CFLs until the full amount of the grants has been transferred to the participants of the project. In the event of misprocurement or any misuse of the grant funds, the World Bank may require the repayment in whole or in part of both grants.

9.2 Conditions for effectiveness

The conditions for effectiveness with respect to the project are: the execution of the trust deeds for the Implementing Units for Guadalajara and Monterrey; the establishment and staffing of the central Project Advisory Committee and the local Technical Committees, and appointment of local managers, with qualified personnel satisfactory to the World Bank; and the execution of the Norwegian grant agreement.

9.3 Environmental aspects

As a signatory of the FCCC, Mexico is committed to contribute to its objective; this project will be a positive step in that direction. Due to the benefits that it produces for the environment, the project has been classified as Category C. The project will first result in global environmental benefits through a reduction in CO_2 emissions of about 118,000 tons annually, thus producing climate related benefits mostly for Mexico. This figure is likely to increase through the free rider effect (or induced sales) of the project throughout Mexico. Secondly, national environmental benefits will be achieved through reduced emissions of sulphur dioxide (3,000 tons annually), nitrogen oxides (205 tons annually) and particulates. These reductions will improve local air quality and result in positive health effects and less damage to crops, vegetation and buildings. The reduction of emissions from power stations in the project areas will add favourably to the national initiatives to deal with these environmental problems.

9.4 Project benefits and risks

The execution of the project will result in substantial economic benefits to: the project participants who will enjoy a comparable or higher lighting level at reduced cost; and society at large and CFE, which will be able to postpone investments for about 100 MW and save about 169 GWh - or 311,000 barrels of Bunker C and

34,000 tons of coal - annually.[3] The results show that for Mexico the internal rate of return (IRR) exceeds 56% for the various sensitivity tests considered; for CFE alone, the IRR exceeds 32% for all probable events, and for the participants the minimum IRR calculated was over 100%. The project thus has very attractive internal rates of return for all parties involved and the results are very robust even under very pessimistic assumptions.

The main risks associated with the project are, first, that market penetration might be slower than expected, which might result in higher project costs as more costly marketing strategies would have to be pursued. Second, the CFLs might be installed in places where they are used for less than four hours a day on average, as assumed. However, the project remains viable even assuming an average use of only two hours per day. Third, the voltage fluctuations in the Mexican power system might be such as to significantly reduce the lifetime of the CFLs; in order to reduce this risk, the specifications will require that fluctuations of plus or minus 10% have no effect on the lifetime of the lamps (voltage fluctuation regulation is smaller than this in Guadalajara and Monterrey). Finally, the CFLs might fail to reach their projected lifetime, even if there were no excessive voltage fluctuations. Given that the lifetime of the CFLs is expected to extend over several years, CFE will perform standard accelerated tests that will be complemented with adequate supplier guarantees. For its part, CFE will guarantee to programme participants replacement of any CFL that fails within two years of the date of purchase; experience has shown that most defective CFLs will fail within this period of time.

---

[3] For a detailed economic evaluation of the project, see World Bank/GEF, 1994.

References

World Bank, 1992. *Mexico: Environmental Issues Strategy Paper*, August 1992.

World Bank/GEF, 1994. *Mexico, High Efficiency Lighting Pilot Project*, Report No. 12448-ME, 8 March 1994.

26. Joint Implementation

Difficult to implement?

KJELL ROLAND and TORLEIF HAUGLAND*

1 The Concept

From the analysis in this volume on the scope and several economic, legal and institutional aspects of JI, it has become clear that implementing JI will raise several problems that need to be solved. For instance, a problem which has regularly been pointed out is how to estimate the net abatement of a particular project, or what incentives exist for the Annex I and non-Annex I countries.

In this chapter some of these conceptual and practical problems will be discussed based on the considerable experience of ECON Energy with (JI) projects. The authors have participated in World Bank missions concerning the pilot projects of the GEF in Mexico and Poland. These projects, which were initiated to demonstrate the viability of the concept of JI, will be briefly discussed in the next section. Section 3 deals with the problems of measuring the net abatement realized by JI projects, and two side-effects that could undermine the success of JI projects: free riding and conservation rebound. Finally, in section 4 the incentives to report on emission reductions by the participants will be dealt with.

2 The Projects

2.1 The ILUMEX project in Mexico

The target of the ILUMEX project is to distribute 1.7 mln. fluorescent lamps (CFLs) over a two year period as a substitute for incandescent lamps (the project has already been discussed in more detail by Luzuriaga, Chapter 25). This is basically done by subsidizing the introduction of this new technology through financial support for the purchase of the lamps or through the electricity bills. The total cost of the project, which has been estimated at $23 mln., will be financed by grant funds from GEF ($10 mln.), Mexico ($10 mln.) and Norway ($3 mln.). The project aims to reduce

---

* Kjell Roland and Torleif Haugland are associated with ECON Energy, Oslo. Jan Arild Snoen and Erik S. Sørensen have provided valuable comments to an earlier draft. Financial support from the Norwegian Research Council is gratefully acknowledged.

C. J. Jepma (ed.), The Feasibility of Joint Implementation, 359–366.

greenhouse gas (GHG) emissions (CO_2 and methane) and to improve the local air quality through a reduction of the emissions of sulphur and particulates.

A precondition for GEF to spend money on global warming abatement initiatives is that the GEF funds should cover the incremental costs of the project: the cost of a project that the host country would not otherwise have carried out (the additionality criterion, see a.o. Pearce, Chapter 1). However, as the ILUMEX project has demonstrated, it is very difficult to verify whether the contribution by GEF actually covers the incremental costs only. Hence the criteria for GEF participating in JI projects are not easy to operationalize.

Estimates of abatement costs that were made for the ILUMEX project range from a net 'negative cost' (*i.e.* net gain) to a net cost in the order of \$30 per ton of CO_2. The cost estimates are sensitive to the assumptions made about: the economic lifetime of the investments; the future energy prices; and the side-effects that may be relevant to a comprehensive cost-benefit assessment of the project. Such a cost-benefit assessment includes the institutional constraints and the lack of capital available for the national electricity utility. In the low cost estimates local environmental (positive) side-effects are normally taken into consideration, whereas in the higher cost estimates, in which the costs of implementing the project are related to the global warming effects only, these and other local effects are ignored. It may be clear from this that the estimation of the abatement costs of a project depends highly on the sort of convention and the assumptions made in the cost-benefit calculations.

2.2 Coal-to-gas (CTG) conversion in Poland

The CTG project basically consists of two components: an investment programme to finance the conversion of district heating boilers from coal to gas; and a technical assistance programme, which includes training and monitoring of the project after completion.

The project costs are estimated at \$44 mln., which will be financed by GEF (\$25 mln.), Poland (\$18 mln.) and Norway (\$1 mln.). The annual abatement costs amount from \$33 to \$84 per ton of CO_2, *i.e.* the project appears to have a lower profitability than the ILUMEX project. Again, the environmental benefits consist of a substantial reduction in CO_2 emissions and an important local air quality component (reducing sulphur, NO_x and particulates emissions). A positive side-effect of the project may be the speeding up of the diffusion of technology which is conventional in the West, but not very well known in Poland.

3 Net Abatement

3.1 The baseline

This section will focus on the two issues raised in the introduction: the net abatement of JI projects that needs to be demonstrated to credit a third party; and the incentives which arise for the parties involved.

In order to estimate the net abatement resulting from a JI project, the difference between baseline and the emission trajectory after the JI project, needs to be defined. In both Mexico and Poland (as in most non-Annex I and Central and Eastern European countries) there are no national emission projections that could serve as a benchmark in terms of making someone (*i.e.* the host government) responsible for the actual emissions. If the countries in question had adopted a national emission target and had politically committed themselves to it, they would have to take full responsibility for deviations between this baseline and the actual, *ex post* emissions. Since this is not the case, we entirely have to rely on the project specific baseline and estimate of emissions. This is where many problems start.

The abatement potential of the projects described in section 2 was initially calculated from an engineering point of view by simply comparing the emissions intensity (emissions per unit of energy services) of the projects with the respective project specific baselines. This approach, however, is too simple as it does not properly take into account the behaviourial factors that may reduce the effectiveness of the programme in terms of net abatement. There are in particular two types of side-effects that may undermine the project effects: *free riding* and *conservation rebound.* These will be discussed in the next sub-sections.

3.2 Free riding

With respect to JI projects, free riders are those parties that would have invested in measures resulting in emissions reductions anyway, even in absence of the JI project. These investments could have been done either at the same time and in an identical manner, with or without the project, or at a later date. These 'delayed investments' can in principle yield higher emission reductions than the 'early investments' triggered by the JI project, in particular when there is rapid technological progress.

For the two projects discussed in this chapter the free rider share is unknown. There are however reasons to believe that the share could be above 50%, *i.e.* more than 50% of the reduction in GHG emissions recorded as a result of the JI projects would have materialized even without the projects. Our presumption of such a high free rider share is based on estimates of free riding in relation to energy conservation programmes in industrialized countries: Norwegian energy conservation programmes with financial investment support show free rider shares of about 70% and many utilities in the United States apply a free rider share of 30-50% in their evaluation of demand-side management measures.[1]

There are, however, some factors that may point towards a smaller free rider share in the two JI projects concerned, when compared to energy conservation programmes in industrialized countries. Firstly, Poland and Mexico have a shortage of investment funds that preclude investors from making profitable investments. Secondly, subsidized energy conservation investments in industrialized countries are often implemented by investors who are actively seeking government support for their already planned investments, resulting in a large free rider share of such programmes. With respect to the two JI projects in this chapter it is unclear whether the selection criteria make it more difficult for investors to seek support.

Contrary to this, there are other factors that suggest that the free rider share could be larger than those estimated for energy conservation programmes in industrial countries. Two factors are of particular importance: future stringent environmental regulations and possible future price reforms. These will be dealt with below.

National and local environmental regulations

Environmental regulations at the local government level may also contribute to reducing GHG emissions in a way JI projects have already done and may thus 'undermine' the effect of a JI project. In Poland, for example, it is being considered to introduce regulations in heavily polluted areas that would mandate a conversion from coal to gas similar to the fuel switching that has already been undertaken by the GEF pilot CTG project. If such regulations are to be implemented, say in one or two years, the net global benefits from the GEF project (the abatement resulting from the GEF project minus the abatement resulting from the local government regulations) would be substantially lower and those who have benefitted from the GEF funds would get a free ride. An even worse case would arise if the foreign assistance provided by JI funding would represent a rather counterproductive incentive to the Polish government: the assumption that foreign financial support will be forthcoming might be an incentive for the government to abstain from, or delay the implementation of stricter national or local regulations to reduce atmospheric emissions.

---

1 Haugland, 1994; Joskow and Marron, 1992.

Price reforms
Energy prices in both Poland and Mexico are below the economic costs of energy supplies. Price reforms would thus reduce both local air pollution and GHG emissions. If such price reforms are not fully taken into account in the baseline assumptions, calculations of the net abatement of the JI projects will overstate the actual benefit, and some of the investors that benefit from the project support may be considered as free riders.

3.3 Conservation rebound

Another possible 'leakage' in the net abatement of JI projects stems from the so-called conservation rebound effect. As with the free rider effect, conservation rebound is a well-known phenomenon from the literature on energy conservation policy. Rebound means that the investment does not result in a reduction in consumption as theoretically projected, because the efficiency improvement triggers behaviourial changes which partly offset the saving. Typically, an investment in insulation, for example, makes a consumer choose a higher indoor temperature. Hence part of the benefit of the investment (*i.e.* the theoretical saving) will be spent on increased comfort. Therefore, the simple engineering calculation overstates the actual environmental benefits if part of the theoretical energy saving and the ensuing financial saving is being spent on higher energy consumption.

The net abatement calculations of the ILUMEX project, for example, are based on estimates which multiplies the difference in Watt per lamp, between CFLs and incandescent lamps, with the number of lamps and the lifetime of the lamps. This approach probably overstates the net abatement of the project, since it is not unlikely that each lamp may be used more hours every day, or that the light output per lamp (the number of Watts), or the number of sockets in the buildings, will increase. The reason for this is that as a result of the energy efficiency improvement, the electricity bill is substantially lower and some of this saving will be used for additional energy services.

Although only a few estimates of the rebound effect exist, from some studies on the rebound in energy conservation programmes an indication of its size can be derived. A survey conducted in 1990 in relation to a Norwegian investment support programme detected conservation rebound in the order of 40% for households and 10% for commerce.[2] An American econometric study of a residential conservation

2 See Haugland, 1994, and Energidata, 1991.

programme found that approximately two-thirds of the initial saving due to engineering effects were eventually eroded by the rebound effect.[3]

It is difficult to judge what the size of the rebound effect will be in the two GEF pilot projects. However, it should be noted that the rebound effect would be high if energy consumers have initially been restrained from having their demand met, which is quite common in the case of electricity demand in developing countries. Hence the rebound effect for certain categories of energy demand could be high and in some case be as high as 100%, *i.e.* the entire theoretical saving is 'eroded' through additional consumption.

3.4 Net abatement at the project level

From the above analysis it can be concluded that both from a conceptual and practical point of view, the net abatement calculations at project level are very complex. It is also worth mentioning that assessing net emissions effects of particular projects, of which the benefits are to be credited to a company or another country, might result in high transaction costs. These can easily outweigh the net benefits of potentially beneficial projects. The only way in which this problem can be met is to define classes of projects for which more general guidelines can be established. In that case, the rationale should be that the estimated emissions reductions are correct on the average but not necessarily in each specific case. To document each specific case in great detail, transaction costs would probably become very high.

4 Incentives to Overreport Emission Reductions

Getting the incentives right for both buyers and sellers to report accurately on net emission effects is crucial in order to improve the credibility of JI. If we start exercising JI now, crediting will have to be the final target after what might be a long trial and error phase. The important challenge is how to make the participants responsible for the actual *ex post* emission abatement effects, instead of for the on paper *ex ante* emissions. Another question is how the risks of project failure in terms of not delivering what was promised should be internalized. The risk dealt with here is not specifically the financial risk that the private sector or foreign governments involved obviously would have to face, but the global risk that a project results in a serious leakage.

3 Khazzoom, 1986.

It seems to be important to distinguish between countries that have a national commitment to keep emissions below a certain level on the one hand and those without such a commitment on the other. Suppose, for example, that the assessment of a particular project is wrong or that unexpected indirect effects occur. This does not really matter to the global environment, if such a project is set up between countries that have a national emission target. If the host country which exports the credit and receives the money miscalculates the abatement effect from the JI effect, it will, because of its national baseline commitment, still be responsible for the emissions originating from its territory. The risk to the global environment will be internalized within the agreement between the governments and the private sector involved and will not be left to a third non-participating party.

For Central and Eastern European or developing countries without national emission targets, the whole effect of the project hinges on the project assessment of emissions. Then the problem of perverse incentives may arise: all the parties involved would like to exaggerate their emissions reductions. For this reason, crediting JI projects in countries without national emission targets would require a strong verification and monitoring mechanism, *i.e.* an independent monitoring and certifying body - a new international bureaucracy. Without such a mechanism crediting of abatement through JI will be a hard task to fulfil.

References

Energidata, 1991. *Resultatmål av ENØK innsats (Evaluating Energy Conservation Policy)*, ED 91-145, Energidata a.s, Trondheim, Norway.

Haugland, T., 1994. *The Cost-Efficiency of Financial Investment Support in Energy Conservation Policy*, EED-working paper no. 6/94.

Joskow, P.L. and D.B. Marron, 1992. 'What does a megawatt really cost? Evidence from Utility Conservation Programmes'. *The Energy Journal*, Vol. 13, No. 4.

Khazzoom, J. D., 1986. *An Econometric Model Integrating Conservation Measures in the Estimation of the Residential Demand for Electricity.* Greenwich, Conn.: JAI Press.

27. Opportunities for Joint Implementation Projects Outside of International Agreements on Greenhouse Gas Reduction

AJAY K. SANGHI, ANTHONY L. JOSEPH and KARL MICHAEL*

1 Introduction

This volume has primarily focused on JI as a cooperative implementation of greenhouse gas (GHG) reduction policies and measures, whereby a party in one country assists a party in another country to achieve an underlying GHG reduction objective. JI projects will only come about if individual countries require mandatory reductions in GHGs by their industries as a result of international agreements. However, opportunities for JI projects could arise outside of an international agreement on GHG reductions, for example as a result of actions taken by individual states in the USA under their environmental and energy regulations. It is likely that potential sellers of GHG reduction credits in other countries are unaware of JI opportunities which may arise from actions of individual states within the USA. An understanding of such opportunities will enhance the development of a market for JI.

This chapter discusses environmental regulations and energy policies being considered or adopted by states of the USA (independent of US/International agreements on GHG reductions) which may give rise to opportunities for JI projects. The second section provides a discussion of a New York State environmental law, which could create opportunities for JI projects. This is followed by a section which discusses New York State requirements to consider environmental externalities in energy decision making, and how these actions could also create opportunities for JI projects. The paper concludes by showing that JI opportunities which may be created by the action of individual states in the USA, outside of international agreements on reductions of GHGs, are substantial.

* The authors are associated with the New York State Energy Office, New York, USA. The views and opinions expressed in this chapter are solely those of the authors and do not necessarily state or reflect views, opinions or policies of the New York State Energy Office.

C. J. Jepma (ed.), The Feasibility of Joint Implementation, 367–377.

2 Environmental Regulations

In the USA, emissions of CO_2 are not regulated under federal and/or state environmental laws. However, some states have implemented environmental regulations which can indirectly address GHG emissions. For example, New York State promulgated a comprehensive law in 1975 known as the State Environmental Quality Review Act (SEQRA). SEQRA provides a process for considering the views of citizen action groups in issuing environmental permits. As a result of such public involvement, SEQRA can impose environmental conditions which go beyond the scope of existing regulations. SEQRA includes the following declaration of purpose:

> It is the purpose of this act to declare a state policy which will encourage productive and enjoyable harmony between man and his environment; to promote efforts which will prevent or eliminate damage to the environment and enhance human and community resources; and to enrich the understanding of the ecological systems, natural, human and community resources important to the people of the state.

Furthermore:

> It is the intent of the legislature that all agencies conduct their affairs with an awareness that they are stewards of the air, water, land, and living resources, and that they have an obligation to protect the environment for the use and enjoyment of this and all future generations.
>
> It is the intent of the legislature that all agencies which regulate activities of individuals, corporations, and public agencies which are found to affect the quality of the environment shall regulate such activities so that due consideration is given to preventing environmental damage.

The SEQRA process requires that an Environmental Impact Statement (EIS) be prepared, subject to review by public hearings and comments. The EIS identifies ways to minimize or mitigate adverse environmental effects, or identifies a potentially less damaging alternative. SEQRA also provides New York State with authority to require mitigation measures for any type of environmental impact, even if there are no specific regulations which address a particular pollutant, such as CO_2.

The SEQRA applies to all construction projects of certain size (except electricity generating sources larger than 80 MW) and to an action which is undertaken or approved by local, regional or state agencies. For purposes of siting electric generating plants which exceed 80 MW, SEQRA is superseded by the 'Facility Siting' process under Article VIII of the New York Energy Service Law. Article VIII was developed as a 'one-stop shopping' approach to the siting and permitting of large power plants. The language in Article VIII is similar to SEQRA with respect to consideration of environmental impacts (*i.e.*, there is authority to require mitigation measures for any type of environmental impacts, within the boundaries of 'reasonableness', even if there are no specific regulations which address a particular pollutant, such as CO_2).[1]

The broad authority of Article VIII was recently exercised in the case of a 210 MW coal-fired cogeneration plant which was proposed to be built by Inter-Power at Halfmoon, New York, near Albany. The Inter-Power project has been strongly opposed by local citizen action groups and national environmental groups such as the Sierra Club, as well as by the neighbouring state of Massachusetts. Based on environmental concerns, these groups have been fighting for either no plant or one fired by natural gas instead of coal.[2]

As a result of the Article VIII process, New York's Siting Board ordered that the proposed plant meet a stringent set of environmental requirements, some of which go beyond the specific requirements of existing environmental regulations, such as restrictions on noise levels and monitoring of toxics in coal and limestone used at the plant. Most importantly, the Siting Board required Inter-Power to implement measures to partially offset its plant's CO_2 emissions. This requirement is in direct response to public concerns that CO_2 emissions contribute to environmental damage by global warming. The Siting Board set the required CO_2 offset expenditure at $500,000 (in 1991 $) per year for a total of $15 mln. over the projected 30-year life of the proposed plant. The CO_2 mitigation measures could include tree planting, energy efficiency or other measures. Inter-power must submit a plan of its expenditures for reducing CO_2 to New York's regulatory agencies each year. The annual

---

1 Note that Article VIII was a temporary law which has technically expired. In 1992, New York passed a new energy law which specifies that a process similar to Article VIII is to be defined by new regulations, but this has not yet been done. In the meantime, the consensus among state regulators seems to be to proceed as if Article VIII is still in effect, as such action is consistent with the intent of the new energy bill.

2 Due to continuing opposition to the Inter-Power plant, final permitting of the proposed facility is still tied up in New York's energy and environmental regulatory process. At this point in time, it remains uncertain whether or not the plant will actually be built.

regulatory review of Inter-Power's CO_2 offset spending is intended to ensure that the tonnage of CO_2 offsets achieved is maximized given the annual expenditure limit.[3]

The CO_2 offset requirement imposed on Inter-Power by New York's Siting Board could result in the implementation of a JI project. For this to occur, the JI project must be proposed by Inter-Power in its annual CO_2 offset plan submitted to New York's regulators. The project could be implemented provided the plan is approved by the regulators as the most cost-effective approach to meeting the CO_2 offset requirement.

3 Environmental Externalities

3.1 Addressing and accounting environmental externalities by private and public sectors within States

The requirement to consider environmental externalities in energy planning and decision making is another state action which could give rise to JI opportunities. Since the 1930s, economists have purported that gains in economic efficiency and resource allocation can be achieved by internalizing externalities.[4] That is, social welfare can be improved by including externalities in market prices. New York State energy planning legislation recognizes the importance of considering externalities in energy planning. It requires that:

> The state energy plan shall provide guidance for energy related decisions to be made by the public and private sector within the state.[5]

The law is broad in its scope. It specifically requires that the energy plan gives 'due regard' to 'adverse and beneficial environmental impacts' in drafting its statement of energy policies and long-range planning objectives and that it provides strategies appropriate to achieve the least cost method of meeting the energy needs of the state.[6] Clearly, to the extent that the plan considers the long range implications of

---

3 By employing relatively low-cost CO_2 reduction measures (i.e., $5/ton of carbon removed), which include increased fuel efficiency for cars and trucks, DSM measures and tropical reforestation, the required annual expenditure could mitigate about 20% of the plant's annual CO_2 emissions.

4 Most notably, Pigou, 1932; Coase, 1960.

5 New York State Energy Office, Sec. 6-104 (3) (a).

6 New York State Energy Office, Sec. 1-104 (2) (d).

residual emissions from energy resources, it must address externalities because the effects of currently legal residual emissions are external to the cost of production and utilization of energy resources.

In addition to the New York's energy planning legislation, the New York State Public Service Commission (NYSPSC) requires the State's regulated electric utilities to account for environmental externalities in electricity resource procurement activities. Currently, New York State electric utilities are required to procure future capacity through a competitive bidding programme. In evaluation of these bids externalities are specifically considered (bids solicited during 1989 through 1991).[7] The utilities have been directed to use monetary values ($/ton) for SO_2, NO_x, CO_2 for such purposes.

Requirements to consider externalities has been adopted by several other state regulatory utility commissions and by state energy planning organizations in evaluating electricity resources. To date, 26 states require their electric utilities to consider air emission externalities in resource decisions.[8] Overall, the requirements are primarily concerned with addressing external costs associated with emissions of air pollutants, particularly SO_2, NO_x, CO_2, VOC/ROG and PM_{10}. Half of the states (thirteen) require externalities to be considered only qualitatively by their state utilities but the other half of the states requires that the externalities be considered quantitatively by appending an externality adder to the market price of the competing resources.

3.2 Use of externality adders in energy planning and decision making

In energy planning or resource selection, the externality adder associated with a policy action or resource is derived as the product of the tonnage emissions of a particular pollutant of the action and the estimated monetary externality value ($/ton) for the pollutant, summed across all pollutants being considered. The resultant externality adder is then added to the cost of the action to derive the total social cost.

---

[7] Competitive bidding by New York State's IOUs (Investor Owned Utilities; these are New York State's seven regulated utilities) was implemented under New York Public Service Commission Order No. 89-15, May 23, 1989. The competitive bidding process allows small power production facilities, cogeneration facilities, other independent power producers, utilities and other wholesale power producers to submit a proposal to sell electric capacity to the host utility under a long-term contractual arrangement.

[8] The number of states may have changed as of the printing of this volume.

Table 27.1 illustrates an application of externality adders in the selection of an electricity generating resource, assuming that residual emissions of CO_2 are the only externality of concern, and a CO_2 externality value of $8.6/ton.[9] The resource choices are hypothetical natural gas-fired and coal-fired electricity generating plants. As shown, the natural gas plant (at 8 ¢/kWh) is a more expensive resource choice than the coal plant (at 7.5 ¢/kWh), based on cost-of-service rates (production costs). However, when each plant's CO_2 externality adder is considered, the natural gas plant becomes a less expensive resource choice than the coal plant.

Table 27.1. Use of externality adder in resource selection: comparison of 200 MW Natural Gas Plant and a Coal Plant[a].

	Natural Gas Plant	Coal Plant
Cost-of-Service Electricity Rate	8.0 ¢/kWh	7.5 ¢/kWh
Annual Residual CO_2 Emissions	540,000 tons	1,470,000 tons
CO_2 Externality Value	$8.6/ton	$8.6/ton
CO_2 Externality Adder	0.4 ¢/kWh	1.0 ¢/kWh
Societal Cost	8.4 ¢/kWh	8.5 ¢/kWh
Resource Choice With Externality Adder		Natural Gas Plant
Annual Reduction in Residual CO_2 Emissions over the Choice of the Coal Plant		930,000 tons
Cost of CO_2 Reduction ($/ton)[b]		$6.6/ton

a Based upon an assumed capacity factor of 70%.
b Total reduction cost is based upon the difference in the annual cost-of service between the natural gas plant and the coal plant.
Source: New York State Energy Office, 1994.

On the surface, the use of externality adders appears to be capable of producing pollution reductions similar to those achieved by use of a pollution tax. However, the externality adder is not actually paid by the polluting resource (and thus ratepayers/consumers) - it is only used in the selection of new resources. Therefore, externality adders generally will not produce the same level of impacts (pollution reductions) as those which would be brought about by a tax equal in size to the underlying $/ton externality value. In some cases the externality adder will not produce pollution reductions at all (unlike a pollution tax/fee). Also, the adder

[9] The CO_2 externality value of $8.6/ton (1992$) is reported in the recently completed *Draft 1994 New York State Energy Plan*, New York State Energy Office, New York State Department of Environmental Conservation, and the New York State Department of Public Services, 1994.

provides no flexibility to pursue lower cost emission reduction possibilities elsewhere in the system through offsets.

3.3 Alternatives to the use of externality adders

The 1994 New York State Energy Plan explicitly recognizes limitations in the use of externality adders to address environmental externality concerns, and recommends the use of alternatives to adders.[10] The underlying objectives of the alternatives are: to encourage actions which reduce residual emissions; and to provide flexibility so that such reductions are achieved at a least cost. The philosophy behind the alternatives is to allow a utility to choose a resource without using externality adders, but require the utility to mitigate a portion of the emissions of the chosen resource through investments in the system and/or through JI projects. Because alternatives are evaluated within a least cost framework, society and electricity ratepayers are likely to be better off than under the rigid system of externality adders.

One alternative suggested in the *1994 New York State Energy Plan* is achievement of a direct environmental improvement target. Another alternative is investment in environmentally beneficial projects.[11] The concept of environmentally beneficial projects embraces the principle that even though direct mitigation of pollution or associated damage may be difficult, there are projects which may be considered beneficial to the environment, and investments in such projects should fulfil the utilities obligation to include externality considerations in energy decision making.[12]

Alternative approaches to adders should be designed so that they can be implemented within existing authority under which state energy planning and regulatory agencies use environmental adders, without further legislative approval. This requires that

---

[10] See note 9.

[11] The idea of using environmentally beneficial projects is borrowed from the early 1980s, when several environmental groups settled Clean Water Act citizen suits with agreements requiring the defendant to perform a significant environmental project, in lieu of civil penalties.

[12] For instance, from 1983 to 1989, New Jersey Public Interest Research Group reached agreements with Clean Water Act (CWA) defendants, approved by the US District Courts, in which $3.3 million was to be spent on environmental projects. Although these projects were environmentally and regionally related to the defendant, they were not directly equated to the particular harm created by the defendants pollution.

In the context of reducing GHG, the concept of investment in environmentally beneficial project could include expenditures on research projects related to GHG reduction measures (e.g., animal waste disposal, efficiency improvement, population control, etc.).

there be an 'equivalence' between use of an externality adder and use of an alternative to it.

The concept of equivalence is illustrated using the example in Table 27.1. As shown, consideration of the CO_2 externality adder changes the least cost resource choice from the coal plant to the natural gas plant. This decision represents an annual CO_2 reduction of about 930,000 tons at an average cost of \$6.6/ton. Thus, an 'equivalent' direct reduction target of 930,000 of CO_2 could be allowed as an alternative to the use of adders. The availability of this alternative would give the utility flexibility to either use adders and choose the natural gas plant, or choose the coal plant and reduce the equivalent level of CO_2 emissions. If sufficient CO_2 reduction actions were available to the utility at under \$6.6/ton, the direct reduction alternative provides opportunity to produce the same level of environmental benefits brought about by use of the adder approach at a lower cost.

Experience shows that the effect on system wide emissions brought about by the use of an externality adder tends to be small and is difficult to estimate (the above example is atypical).[13] Thus, it is not easy to establish an 'exact' equivalence between use of an externality adder and use of an alternative approach. Given these considerations, it is necessary to have generic rules for establishing equivalence.

The imputed externality cost associated with a utility decision (the product of the resultant emissions times the corresponding externality value) provides a reference point for establishing equivalence between externality adders and alternatives to externality adders. That is, the total imputed externality cost provides a basis for defining a direct utility expenditure on pollution reduction and/or environmentally beneficial projects considered to be an adequate alternative to the use of adders.

In using the total imputed externality cost to specify equivalence, it is important to recognize that use of an adder will not result in a utility incurring expenditures anywhere near levels indicated by this cost (because they are not imposed as pollution fees). Thus, the equivalent expenditure must be set considerably lower than

[13] If resource options have the same emission characteristics, externality adders will have little or no impact on the least cost ranking of the resources, compared with the ranking without consideration of adders. This does not imply that the resource choice is environmentally acceptable, and that no further pollution abatement is required. In New York State for example, most of the electricity replacement capacity in the foreseeable future is expected to be fuelled by natural gas. Simply because a technology such as natural gas generation is economically competitive, does not suggest that environmental externalities are internalized and that environmental concerns of energy planning are automatically satisfied. In other words, the condition that the externality adder does not change the order of selection of proposed resources does not imply that environmental externalities have been fully internalized (i.e., that there is no obligation to meet the environmental objective underlying the externality value).

the imputed externality cost, if the utility is to perceive making such an expenditure as a viable alternative to the use of adders. It is suggested that the equivalent expenditure be set at 10% of the imputed externality costs.

Equivalence should also be specified as a tonnage reduction alternative. This provides flexibility to reduce the cost of considering the externality (to the extent it can purchase offsets at a cost which is lower than its average control cost). Thus, it is suggested that a utility be allowed to reduce or offset the resource's emissions by 10% as an alternative to the use of adders. For example, the imputed CO_2 externality costs for the coal plant in Table 27.1 is estimated at about \$12.6 mln. per year (1.47 mln. tons of CO_2 times an externality value of \$8.6/ton). Based on the 10% rule suggested above, the alternatives to adders include: making an expenditure (equivalent) of about \$1.26 mln. annually (for direct investment in pollution reduction, including JI and environmentally beneficial projects); or offsetting CO_2 emissions by about 0.15 mln. tons annually. This allows the utility to choose the annual offset alternative of 0.15 mln. tons of CO_2, rather than the \$1.26 mln. per year expenditure alternative, if it can achieve the CO_2 reduction target by utilizing low cost options and investing in JI projects at a cost of less than \$1.26 mln. per year.

4 Discussion

Besides New York, many other states in the USA have their own environmental laws or administrative/regulatory requirements which go beyond implementing federal environmental laws. Under various state requirements, utilities are being held responsible for damages from their residual emissions. This is particularly being done in the choice of new electricity resources.

The Wisconsin Public Service Commission has recently adopted regulation which provides an alternative to the use of the state's \$17/ton CO_2 externality adder.[14] Wisconsin utilities are allowed to reduce the number of tons of CO_2 affected by the adder, by offsetting CO_2 emissions through use of efficiency measures, renewable resources and purchase of offsets. The tons of such CO_2 offsets are used to reduce the tons of CO_2 applied to the adder in the ratio of one to one.

The Wisconsin rule represents a step in the direction of offering offsets as an alternative approach to the use of adders and in formally recognizing that adders are not to be applied to CO_2 emissions which have been voluntarily offset. California has used a risk shifting mandate. Under California's rule new long-term energy supply

[14] Wisconsin PSC Order, 1994. Docket No. 05-EP-6.

contracts must require the suppliers to assume the full risk of future CO_2 emission limitations, taxes or regulations. This clause can be analyzed as a type of government-imposed insurance policy. By requiring the long-term supplier to bear the entire cost of any future carbon emission regulations, taxes or charges, the supplier becomes the insurer against any adverse economic consequences of future carbon regulations.[15]

This chapter has presented some examples of how state actions could create additional markets for JI projects irrespective of the federal action. State actions such as the New York Siting Board decision to require CO_2 offsets for the proposed Inter-Power plant could create substantial markets for JI projects. A conservative estimation suggests that the market could reach a cumulative value of more than $1 bln. in next 15 to 20 years.[16] Note that, even expenditures of a few hundred thousand dollars per year from an individual case, such as the Inter-Power plant, could sustain a viable JI project in a developing country.

Parties interested in JI should keep abreast with developments in the different states of the USA by keeping in touch with state utility regulatory commissions, state energy planning agencies, and electric utilities. It will be useful for them to establish good communications with state agencies by letting them know about their projects. Also, potential suppliers of GHG reduction credits should attempt to understand and address concerns, such as verification and sustainability of the reduction, in allowing participation in JI projects to satisfy their state mandates. This cooperative effort and understanding will clear the way for a quick action at the appropriate time.

---

[15] Without this rule, utility customers become the 'unwitting and uncompensated insurers of the fossil fuel industry,' which has not offered to indemnify the public against the risks (Cavanagh, 1993).

[16] The cumulative estimate of $1.2 billion is based on the following assumptions: DOE Reference Case forecast of 172 GW of new fossil fuel-fired electric generation capacity by 2010 (*Annual Energy Outlook 1994*); 70% gas-fired and 30% coal-fired; thirteen states apply some monetary value of CO_2 emissions in selecting new generation resources; 10% of the CO_2 emissions from new capacity in those thirteen states is required to be offset or an equivalent expenditure made in the amount of $9 per ton of CO_2 (median value of states where CO_2 externality value is now considered in decision-making); using the above methodology, the cumulative value of the JI market could range from about $135 million (based on the New York Public Service commission value of $1 per ton of CO_2) to about $3.0 billion (based on the $22 per ton value used by Massachusetts and Nevada).

References

Cavanagh, 1993. 'Response to a Letter to the Editor', 6 *Electricity Journal*, 3 May 1993.
Coase, R., 1960. 'The Problem of Social Costs', *Journal of Law and Economics*, Vol. 3.
New York State, 1975. *State Environmental Quality Review Act (SEQRA)*, New York, USA.
New York State Energy Office, Sec. 1-104 (2) (d).
New York State Energy Office, Sec. 6-104 (3) (a).
New York State Energy Office, New York State Department of Environmental Conservation, and the New York State Department of Public Services, 1994. *Draft 1994 New York State Energy Plan*, 'Environmental Externalities', Volume II, Issue 8, February 1994.
Pigou, A.C., 1932. *The Economics of Welfare*, Macmillan, London.
US Department of Energy/Energy Information Administration, 1994. *Annual Energy Outlook 1994: With Projections to 2010.* DOE/EIA-0383(94), Distribution Category UC-950, January 1994.
Wisconsin PSC Order, 1994. Docket No. 05-EP-6.

Concluding Address

GERARD WOLTERS*

As Minister Alders noted in his opening address, no one at UNCED could imagine to the full extent what it meant to include the option of JI in the text of the FCCC. During the Groningen Conference the concept of JI has been considered from a lot of different angles. It was for instance seen as a new challenge for cooperation between developed and developing countries thereby investing in world peace and security. Moreover, it has been recognized as a promising instrument which could make a significant contribution to achieving the long-term goal of the FCCC. We have looked not only at the opportunities JI offers, but also at the risks it entails. We have discussed ways to maximize the benefits and minimize the risks.

It has been stressed over and over again that incentives play a crucial role. Since we wish to achieve the ultimate goal of the FCCC, we are interested in incentives both for States - especially the developed countries - to take on progressively stricter commitments, and for firms, various social organizations, and provincial and municipal governments to get involved in JI. In order to properly assess the role that JI can play it was generally felt also at the Conference that a period of gaining experience is indispensable.

From an economic point of view it was emphasized that the practical problems that have been signalled so far are only worthwhile solving if the potential cost savings of JI are indeed as large as we think they are. Therefore, this should be a major priority for research.

A very interesting idea for a phased approach was presented, zooming in on costs and benefits at different levels. It would allow the Parties to the FCCC to gradually develop a mature system, while fruits can already be reaped soon. However, some form of recognition was thought to be essential in order to encourage the private sector to make a start. The latter point was underlined from the side of the business community. Very interesting pilot projects and potential pilot projects have been presented. It was argued that it should not really matter whether JI projects make a profit, since profits are also made on domestic reduction measures. Others held different views. I think it is a matter for serious debate whether or not the Parties to the FCCC should restrict JI to projects which have net costs and leave the profitable projects to the commercial sector. The cost curves in this volume have demonstrated that we will move beyond the no-regret phase soon! It is clear that the relation

* Gerard J.R. Wolters is Deputy Director-General for Environment of The Netherlands' Ministry of Housing, Spatial Planning and Environment.

C. J. Jepma (ed.), The Feasibility of Joint Implementation, 379–380.

between the profitability argument and the additionality and incentive requirements needs further analysis and discussion in the framework of the INC, the CoP, and its subsidiary bodies.

The Groningen Statement on Joint Implementation adopted by the participants expresses the general feeling of the Conference on the further development of JI. It stresses among other things that JI should be a supplement to domestic action by developed countries due to the urgency of the global situation and that projects would have to follow objective and verifiable criteria to be established by the CoP. Furthermore, it acknowledges the need for both further research into several unresolved questions and for international cooperation to gain experience with merits and demerits of JI.

Several countries, including The Netherlands are demonstrating that attitude by developing and implementing national pilot programmes. Since the composition of the Conference could be considered fairly representative for the world community the Groningen Statement, coming as it does at this timely moment, will no doubt be received as a valuable input to the discussions in the next two sessions of the INC.

As host country, The Netherlands is glad that the Conference has been attended by experts from so many different countries and backgrounds. We think that it is very important that the discussion has thus been broadened to countries and parts of society that were previously not involved in the subject.

This is what the world needs in order to bring the discussion further, and in fact brings JI closer to actual practice. We have observed a lot of interesting contacts being established during the Conference, and we are confident that their fruits will last far beyond this moment.

Last but not least, you have recommended the organizers to explore the possibilities of efficient ways to sustain the process started by the Conference, such as a JI Newsletter, a network for the exchange of information on JI studies and projects, and follow-up conferences.

The Netherlands' Ministry of Housing, Spatial Planning and Environment hopes and expects that the exchange of views that has taken place, will contribute to many jointly implemented ventures in the future.

Index

MIX
Papier aus verantwortungsvollen Quellen
Paper from responsible sources
FSC® C105338

If you have any concerns about our products,
you can contact us on
ProductSafety@springernature.com

In case Publisher is established outside the EU,
the EU authorized representative is:
Springer Nature Customer Service Center GmbH
Europaplatz 3, 69115 Heidelberg, Germany

Printed by Libri Plureos GmbH
in Hamburg, Germany